ALGEBRA 1-2

LARSON, KANOLD, STIFF

APPLICATIONS HANDBOOK

Copyright © 1993 by D.C. Heath and Company

A Division of Houghton Mifflin Company

All rights reserved. No part of this publication may be reproduced or transmitted in any form or by any means, electronic or mechanical, including photocopy, recording, or any information storage or retrieval system, without permission in writing from the publisher.

Published simultaneously in Canada

Printed in the United States of America

International Standard Book Number: 0-669-29944-8

10-DBH-97 96

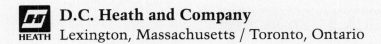 **D.C. Heath and Company**
Lexington, Massachusetts / Toronto, Ontario

CONTENTS

ORBITS

*O*ur universe is in constant motion. Moons revolve around planets. Planets and their moons revolve around stars. The stars and all their satellites travel outward through space at thousands of miles per hour in what seems to be a never-ending hurry to be somewhere else.

A Planet In Motion

595 million miles - that's the distance Earth travels around the sun every year, at an average speed of 66,000 miles per hour. The path Earth takes around the sun is called an **orbit**. This elliptical, or oval-shaped, journey takes about 365 days, 6 hours, 9 minutes, and 10 seconds (a full year) to complete.

The distance from Earth to the sun varies, depending on Earth's position in its orbit. When Earth has reached its closest position to the sun, around January 3, then it has reached **perihelion**. The farthest position of Earth from the sun, called **aphelion**, occurs around July 4.

Our planet, however, does not simply orbit its star. Actually, the yearly journey around the sun is just one of three patterns of motion Earth makes.

In ancient times, people believed that day and night occurred because the sky moved around the world. Today, of course, we know that Earth spins like a top on its **axis**, an imaginary line that cuts through the planet. This daily spinning motion, which takes Earth twenty-four hours to complete, is the planet's second pattern of motion. Day and night, sunrise and sunset are all created by the rotation of Earth on its axis.

While Earth is rotating on its axis and revolving around the sun, it is in the process of yet a third fantastic voyage or pattern of motion. Earth and the rest of the solar system revolve around the center of our galaxy, the Milky Way. The solar system takes about 250 million years to complete this tour.

> **WORD MEANINGS**
> - **orbit**
>   orbis (LATIN)
>   circle or wheel
> - **perihelion**
>   peri (GREEK)
>   around or near
> - **aphelion**
>   apo (GREEK)
>   from or away
>   helios (GREEK)
>   sun

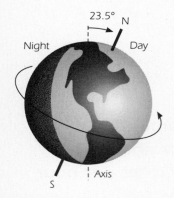

Earth spinning on its axis

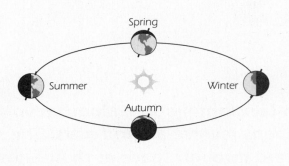

Earth revolving around the sun

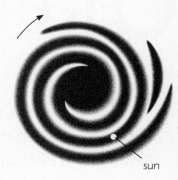

The solar system revolving around the Milky Way Galaxy

A Planet For All Seasons

While standing in a bone-chilling blizzard in Buffalo, New York, on a winter day, it seems hard to imagine that the dog days of summer will arrive again in only a few months.

Have you ever wondered what causes the change of seasons? Although Earth's distance from the sun varies during the year, that is not what causes the change of seasons. Seasonal changes are the result of the tilting of Earth's axis and the position of Earth in its orbit around the sun.

Earth's axis is tilted at approximately 23½° from a line that is vertical to Earth's orbital plane. The axis always points in the same direction. Because the axis is tilted, the northern hemisphere leans away from the sun during one part of Earth's orbit (during the northern winter), and toward the sun during another part of Earth's orbit (the northern summer). The southern hemisphere experiences the exact opposite affect. When the northern hemisphere leans away from the sun, the southern hemisphere leans toward the sun (during the southern summer).

When a place on Earth experiences summer, the rays of the sun fall more directly on that place and heat it more effectively. In the winter, the sun's rays slant more and heat it less effectively. Also, during summer, there are more hours of daylight.

So, when the people living in Buffalo, New York, in the northern hemisphere, know time has come to get out the snow shovels, the people in Buenos Aires, in the southern hemisphere, are putting away their skis and trying on last year's bathing suits.

Solstices

When one hemisphere leans towards the sun, summer occurs. When a hemisphere leans away from the sun, winter occurs. On the first day of summer in the northern hemisphere, precisely at noon, the sun passes through a point in the sky called the **summer solstice**. At that moment, the sun is directly overhead at the Tropic of Cancer. This takes place on either June 20 or June 21. This first day of summer is the longest day of the year in the northern hemisphere, and at the same time, the shortest day of the year in the southern hemisphere.

On the first day of winter in the northern hemisphere, at noon, the sun passes through a point in the sky called the **winter solstice**. At that moment, the sun is directly overhead at the Tropic of Capricorn. This takes place around December 21. This is the shortest day of the year in the northern hemisphere, and also the longest day of the year in the southern hemisphere.

The word solstice is derived from *sol*, which means "sun," and *stice*, which means "stand". The people of ancient times believed that, at the time of solstice, the sun seemed to stand still.

Equinoxes

On the first day of spring in the northern hemisphere, around March 21, at noon, the sun passes through another point in the sky called the **vernal equinox**. At that moment, the sun is directly overhead at the equator, and neither pole leans towards the sun.

The sun is again directly overhead at the equator at noon on the first day of autumn in the northern hemisphere. The point in the sky that the sun passes through at that moment is called the **autumnal equinox**. It occurs around September 23.

FUNTASTIC FACTS

• The universe is so vast that professional astronomers have to depend on amateurs to discover and name some stars.

• The sun is so large that all the planets and moons in our solar system could fit inside it.

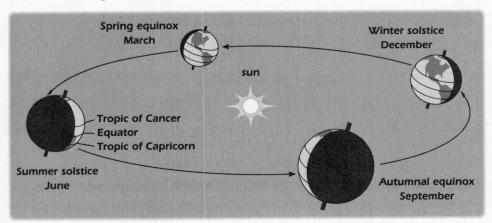

Spring equinox
March

Winter solstice
December

sun

Tropic of Cancer
Equator
Tropic of Capricorn

Summer solstice
June

Autumnal equinox
September

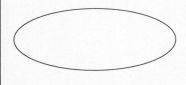

ELLIPSE

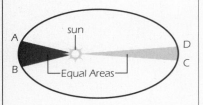

KEPLER'S SECOND LAW
The time to travel from A to B is
the same as from C to D.

$$\left[\frac{x^2}{y^3}\right]$$

Johannes Kepler's Laws

German-born Johannes Kepler was a revolutionary scientist who contradicted the ancient belief that planets travel in circular orbits around the sun. Kepler discovered that the planets revolve around the sun in elliptical orbits; this theory is known as **Kepler's First Law**. According to Kepler, because of elliptical orbits, planets travel closer to the sun at certain times than at others.

Kepler's Second Law states that planets move faster as they approach the sun. Kepler described an imaginary line between the sun and an orbiting planet. The line sweeps across a wide, short area when the planet is nearest the sun and moving at its fastest speed. Similarly, this imaginary line sweeps across a long, narrow area when the planet is farthest from the sun and traveling at its slowest speed. This relationship between a planet's speed and its distance from the sun is also known as the **law of areas**.

Kepler's Third Law states that the time a planet takes to orbit the sun, or **orbital period**, depends on the planet's average distance from the sun, which is measured in **astronomical units (AU's)**. One astronomical unit is the average distance between Earth and the sun, or 93 million miles. Kepler knew that before the average distance from the sun could be determined, the orbital period had to be measured. According to Kepler, the square of the orbital period divided by the cube of the distance from the sun is the same for all planets. In other words, the squares of the orbital periods of two planets are directly proportional to the cubes of their average distances from the sun.

Kepler's knowledge of mathematics led him to the discovery that planetary orbits are elliptical and not circular. His mathematical calculations only worked when an elliptical model was used to describe a planet's distance from the sun and its orbital period.

Kepler's discoveries were crucial to our understanding of the mechanisms of the solar system. When Nicolaus Copernicus proclaimed that Earth was not the center of the universe, a belief that was at least 2000 years old, he was considered a crackpot and a heretic. Kepler was the first astronomer to prove the Copernican theories. Kepler's Laws of Planetary Motion also laid the groundwork for Isaac Newton's discovery of universal gravitation.

Orbits Of Comets

Kepler proved that planets travel around the sun in elliptical orbits. A **comet,** a hunk of rock, ice, dust, or frozen gases that travels through space, is another type of heavenly body that orbits the sun in an elliptical pattern. A comet's **period** is the time it takes

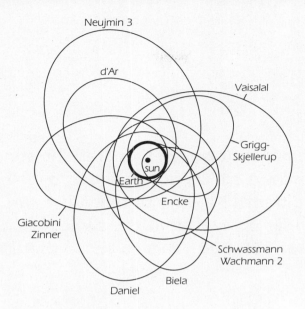

Neujmin 3

d'Ar

Vaisalal

Grigg-
Skjellerup

sun

Earth

Encke

Giacobini
Zinner

Schwassmann
Wachmann 2

Daniel

Biela

HALLEY'S COMET

Halley's Comet rears its head (and tail) on a more or less regular basis...for a comet, that is. The elusive Halley's was first spotted in the heavens as early as 240 B.C. Edmond Halley noted the comet's return in 1682, and correctly predicted its reappearance in 1758. Since then, star-gazers reported sightings of Halley's comet in 1910 and 1986.

for a comet to complete its elliptical orbit. The period of some comets is as short as seven years; other comets are much slower and may not complete their elliptical orbits for thousands, or even millions of years.

ORBITS OF COMETS

Comet	Period of Orbit in Years (approx.)	First Sighted
Halley's Comet	76 to 79	240 B.C.
Biela's Comet	6.75	1772
Encke's Comet	3.3	1786
Comet Faugergues	3000	1811
Comet Pons-Winnecke	5.6 to 6.3	1819
Great Comet of 1843	513	1843
Donati's Comet	2000	1858
Comet Humason	2900	1961
Comet Ikeya-Seki	880	1965
Comet Kohoutek	75,000	1973

FURTHER EXPLORATION

What would happen if Earth suddenly stopped its orbital path around the sun and reversed direction? Explain your answer by drawing a diagram.

STARS

A **star** is a mass of extremely hot gases. Our star, the sun, is about 93 million miles from Earth. A spacecraft traveling over 25,000 miles per hour would take more than 152 days to reach the sun. The next closest star to Earth is so far away that at the same rate, the spacecraft would take about 115,000 years to arrive there.

FUNTASTIC FACTS

▪ If every person in the world were to count the stars, each person could count over 50 billion without naming the same star twice!

▪ Some stars are known as supergiants. The largest of these stars have a diameter about one thousand times that of the sun's.

▪ Astronomers estimate that the universe contains over 200 billion billion stars, although only about 6,000 of them can be seen from Earth with the naked eye.

Star Brightness
The brightness of a star depends on the amount of light energy a star emits, the color of the star, and its surface temperature. Astronomers measure and compare brightness, or luminosity, of stars by using a number called **magnitude**.

The largest stars and those nearest to Earth are not always the brightest. The star Rigel is smaller and farther from Earth than Betelgeuse; but Rigel is much hotter. It is a white star with a surface temperature of 25,000°C, while the closer Betelgeuse is an "M" star, burning orange-red with a surface temperature of 3400°C.

A star's brightness can be measured with an instrument called a **photometer**, which is attached to a telescope. When light from a star enters the photometer, it produces an electric current. The strength of the electric current is measured to determine the star's brightness.

The brightest stars have the lowest magnitudes, and some of the brightest stars have magnitudes so low that their brightness is expressed in negative numbers.

Because they are so great, distances between stars are usually measured in light-years. A **light-year** is the distance light travels in one year. Light travels at about 186,000 miles per second, or about 5.9 trillion miles per year.

Parallax
is an apparent shift in the position of an object when it is viewed from different places. It is a method of measurement that astronomers use to determine the distance from Earth to nearby stars.

★ THE TWENTY BRIGHTEST STARS ★	
Star	Distance in Light-Years from Earth
1. Sirius	8.8
2. Canopus	98
3. Alpha Centauri	4.3
4. Arcturus	36
5. Vega	26
6. Capella	46
7. Rigel	900
8. Procyon	11
9. Betelgeuse	300
10. Achernar	114
11. Beta Centauri	490
12. Altair	16
13. Alpha crucis	370
14. Aldebaran	68
15. Spica	300
16. Anatares	400
17. Pollux	35
18. Fomalhaut	23
19. Deneb	1600
20. Beta Crucis	490

To better understand parallax, hold a pencil at arm's length and look at it with alternate eyes. Line up the pencil with an object in the distance such as a tree. Close one eye and then the other, but don't move the pencil. The pencil will no longer be aligned with the tree. The tree will seem to have shifted because you are looking at it from a slightly different direction. If you knew the distance between your eyes and the angular shift of the pencil, you would be able to determine the distance of the pencil from your face. The apparent shift of the angle of the pencil would be the parallax.

Astronomers use this same method to measure the distance of stars against the background of other remote stars in the universe. Two points on the orbit of Earth which are two astronomical units (about 186 million miles) apart are used as a base line. This is an enormously long base line (especially when compared to the distance between your eyes).

A star's parallax must be observed at various intervals over the course of several months; during this time Earth moves its position between widely separated points in its orbit around the sun. Earth swings about 186 million miles over the course of six months from one side of the sun to the other while making its orbit.

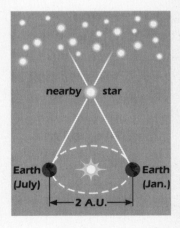

PARALLAX

READ MORE ABOUT ASTRONOMY

• Moore, Patrick. **The Amateur Astronomer.** New York, London: W. W.Norton & Company, 1990.

• Moore, Patrick. **Astronomer's Stars.** New York, London: W. W. Norton & Company, 1987.

• Schwartz, Julius. **Earthwatch.** New York, et al: McGraw-Hill Book Company, 1977.

Other Methods Of Measuring

A **parsec** is a unit of distance used by astronomers that relates to parallax. One parsec is the equivalent of 3.26 light years, or 19 trillion miles. Astronomers can use the parallax method to measure distances of about 300 parsecs.

The parallax method cannot be used to measure stars that are 160 light years away or more because the angular shift is so small it cannot be measured accurately. Another indirect method of finding the distance of very remote stars is to measure their luminosity (actual brightness) and compare these measurements to the luminosity of the sun.

Eclipses

An **eclipse** takes place when one object travels through the shadow of another object. Eclipse is derived from the Greek word, *ek-leipsis*, meaning "forsaking." Early man must have thought the world was coming to an end when, in the middle of the day, the sky suddenly grew dark, and the moon mysteriously passed in front of the sun, casting its shadow on Earth.

Eclipses are caused by the enormous shadows of Earth and the moon. Earth and the moon always cast their shadows into space. The moon orbits Earth about once every month. But a solar or lunar eclipse does not occur every month. One reason is that the moon's orbit is tilted about 5° to that of Earth's orbit around the sun. As a result, a solar eclipse does not occur because the moon's shadow misses Earth most of the time. The moon, as well, misses being eclipsed by moving either above or below Earth's shadow. The only times a solar or lunar eclipse occurs is when Earth, the moon, and the sun are in nearly a straight line.

Solar Eclipses

During a solar eclipse, the moon passes between the sun and Earth, hiding the sun and casting a shadow on a part of Earth that falls in the shadow. Most total solar eclipses last for about 2½ minutes, although some have endured for as long as 7 minutes 40 seconds. A complete darkening of the sun by the moon is called a **total solar eclipse**. A partial darkening of the sun by the moon is called a **partial solar eclipse**. An **annular solar eclipse** occurs when the moon is at its farthest point from Earth so that only the center part of the sun is blocked, leaving a bright ring.

When a total solar eclipse occurs, the **corona**, or outer atmosphere of the sun flashes a brilliant light around the darkened sun, much like a halo. Astronomers have been able to take measurements of the corona during a total solar eclipse.

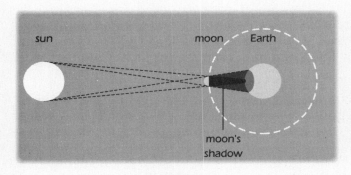

SOLAR ECLIPSE

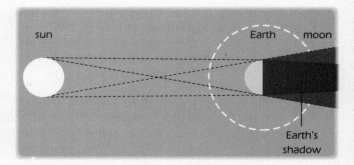

LUNAR ECLIPSE

People living in certain places on Earth can see a total eclipse. This area is called the **path of totality** and is never wider than 170 miles or 274 km.

Lunar Eclipses

A **lunar eclipse** occurs when Earth is directly between the sun and the moon, and the moon moves into Earth's shadow. At that time, people observing the nighttime sky are able to see the moon slowly being covered by shadow.

A **total lunar eclipse** occurs when the entire moon is darkened by Earth's shadow. A **partial lunar eclipse** takes place when a portion of the moon falls into the shadow of Earth. Unlike solar eclipses, lunar eclipses may last for an hour or more.

Eclipses can be predicted with great accuracy. Depending on one's location on Earth, at least two solar and possibly three lunar eclipses may be visible each year.

Eclipses As Learning Tools

Astronomers have learned many things about the universe during eclipses. In 1939, for example, scientists noted that the moon's surface cooled very rapidly during a lunar eclipse. They concluded that the surface of the moon is covered by a fine layer of dust. In the 1960's, a moon-probe and later landings by astronauts proved this theory to be correct.

In 1919, during a solar eclipse, scientists noted that a star that should have been blocked by the sun was clearly visible, showing that its light was being bent. This new information verified the predictions of Einstein's General Theory of Relativity, published in 1916.

Star Energy

The sun is made mostly of hydrogen atoms. The temperature inside the sun is at least 14 million°C. At that temperature there inside the sun, hydrogen atoms are fused together to form helium. Four hydrogen atoms are needed to make one helium atom. Every time this fusion occurs, energy is released and some mass is lost. This is the basic idea behind a fusion reaction. A **fusion reaction** occurs when atoms of one element are fused together to become a larger element. The sun and most stars can be considered fusion reactors.

Scientists today are attempting to duplicate the fusion reaction. However, the extremely high temperatures required for it to occur continues to present a problem. Once solved though, fusion reactors may supply mankind with a cheap, pollution-free energy source.

FURTHER EXPLORATION

Work in groups to discuss why the sun appears to rise in the east and set in the west.

JUPITER

*J*upiter, the largest planet in the solar system
and the fifth planet from the sun, is about 1000 times larger
than Earth. In fact, Jupiter is so large that if it were hollow,
all the other planets in our solar system could be poured inside it,
and room would still be left over.

What Jupiter Looks Like

Jupiter is a stormy world, with clouds that change by the hour in shape, size, and color. Jupiter is sometimes referred to as a fluid planet. Liquid hydrogen and helium is thought to surround a small amount of rocky, iron-like material at its core. In viewing the planet through a telescope, Jupiter appears as a yellowish disk with a series of **belts** and **zones.** Belts are dark lines that circle Jupiter, parallel to its equator. Zones are light-colored areas between the belts and are made of ammonia and water crystals. These belts and zones are caused by Jupiter's swift rotation on its axis.

Voyager 1, a 1978 space-probe, discovered that wrapped around Jupiter is a single ring, a thin veil of dust that appears to be floating on a beam of light. Astronomers speculate that the ring may be composed of debris from comets and meteoroids.

Jupiter has seventeen or more moons. Its four largest moons are called the "Galilean satellites" because they were discovered by the astronomer Galileo in 1610.

The Great Red Spot

Astronomers refer to **The Great Red Spot** as the "eye of Jupiter." This intense atmospheric disturbance is big enough to swallow Earth. The Great Red Spot's hurricane-like winds create a churning sea of many beautiful colors. It is located in Jupiter's southern hemisphere near the equator, and measures more than three times the diameter of Earth. Astronomers have observed that the vivid colors of this unusual storm sometimes appear dimmer than at other times.

FUNTASTIC FACTS

▪ Are the radio noises that astronomers have detected from Jupiter being broadcast by other life-forms operating a radio station? Actually, all the ruckus is the result of the disturbances created by Jupiter's many storms that cause radio energy to be received on Earth.

Jupiter's Rotation And Orbit

Earth's orbit around the sun takes 365 days; Jupiter's orbit takes the equivalent of 12 Earth years to complete. What Jupiter lacks in orbital speed around the sun, it makes up in rotational speed on its axis. Jupiter's rotational speed at its equator is about 45,000 km per hour (or about 28,000 miles per hour). Earth's rotational speed is only about 1600 km per hour (or approximately 995 miles per hour).

Jupiter spins faster than any other planet, rotating on its axis every 9 hours 55 minutes. The rapid spinning of the planet causes Jupiter to bulge at its equator and flatten at its poles.

The Make-Up Of Jupiter

Jupiter is like a frozen star. An icy frosting of clouds creates a cold atmosphere near the surface of Jupiter; but far into the planet's core, astronomers believe that the temperature could be as hot as 24,000°C. This combination of extremely hot temperatures below the surface and extremely cold temperatures above it makes Jupiter a veritable pressure-cooker. The heat rising from the planet in conjunction with Jupiter's fast spinning motion are thought to cause the violent storms in the atmosphere. Jupiter's atmosphere is approximately 88% hydrogen, 11% helium, and 1% ammonia, methane, and water.

CAREERS IN ASTRONOMY

If you wish to pursue a career in astronomy, you have a choice of specializing in many different fields. Stellar astronomers research the composition of stars and how they create light. Solar astronomers specialize in the study of the sun, while planetary astronomers devote their attentions to planetary conditions. The history of the universe and its structure are studied by astronomers called cosmologists. Most astronomers, regardless of their special area of research, are also astrophysicists because they study chemical and physical processes in the vast universe.

Students who are interested in astronomy as a profession should take as many mathematics and physical science courses as possible both in high school and college. Chemistry, biology, and computer science are also important. Graduate work is an integral part of an astronomer's education; most astronomers eventually earn Ph.D degrees.

Professional astronomers teach in universities and conduct research in institutions, national observatories, or laboratories often funded by the federal government. Some work in industry, monitoring pollution, while others are employed by planetariums.

FURTHER EXPLORATION

Work in groups to research what makes Jupiter different from Earth and from the other planets in the solar system. Find out what role gravity plays in making Jupiter a unique planet.

PUNNET SQUARES

*A*re you left-handed or right-handed? Is your hair blond or brown?
Do you have blue eyes or brown eyes?
Inherited traits such as these help make you unique.

Traits, Genes, and Alleles

Traits are inherited characteristics that are carried from one generation to the next. Characteristics of heredity are determined by distinct units called **genes**. For each characteristic, such as left-handedness, a person carries two alternate forms of a gene, one inherited from each parent. These alternate forms of a gene are called **alleles**. Each parent passes one allele for each trait to his or her offspring.

Two factors control any single trait. A **dominant** factor, such as right-handedness, is the stronger of the two and will appear more often as a characteristic. A **recessive** factor, such as left-handedness, is masked in the presence of a dominant factor.

Geneticists use capital and lower case letters to represent alleles. Dominant alleles are represented by capital letters and recessive alleles are represented by lower case letters. Curly hair is a dominant allele and is represented by *C*. Straight hair is recessive and is represented by *c*.

Combinations Of Alleles

Living things receive two alleles for each trait, one from each parent. So, for each trait two letters are needed to illustrate each allele. This combination of alleles, one received from each parent, is called a **genotype**. Suppose you inherit a curly hair allele from your father and a straight hair allele from your mother. The genotype for this trait would be *Cc*.

What other combinations of alleles are possible? A combination of two alleles that each carry the recessive gene for straight hair would have the genotype *cc*. Two curly hair alleles would have the genotype *CC*. Genotypes with identical alleles such as *CC* and *cc* are called **homozygous**, while those with different alleles, such as *Cc*, are called **heterozygous**.

FUNTASTIC FACTS

- Did you know that the ability to roll your tongue is a dominant trait? Try to roll your tongue and see if you are a "roller" or a "non-roller."

- Look in the mirror. If your ear lobes are attached you have inherited a recessive gene.

A **phenotype** is the way in which a genotype is expressed. Since curly hair is a dominant allele, a person carrying the genotype *CC* or *Cc* will have the phenotype of curly hair. Even though the phenotype for these two genes are the same, the genotypes of *CC* and *Cc* are different. To acquire the straight hair phenotype, the genotype would have to be *cc*.

Using Punnett Squares
It is possible to predict the phenotype (or how a genotype will be expressed) for any given trait controlled by simple dominance. The **Punnett square** is a chart based on mathematical probability or chance used in predicting all possible combinations of any phenotype. The square was developed by R.C. Punnett, a biologist and mathematician.

The Punnett square on the left illustrates a cross between two pea plants in which both sperm and egg are heterozygous for height. The allele for tallness is written *T*. The allele for shortness is written *t*. The alleles that are not circled illustrate the new combinations of genotypes.

Notice that one box contains a genotype of *TT*: the phenotype will be tallness. Two boxes contain the genotype *Tt*; since tallness is dominant, the phenotype for both of these boxes will be tallness. One box contains a genotype of *tt*. In that case, shortness will be the phenotype.

When the combination of these genotypes is expressed as a mathematical ratio, it looks like this: 1*TT*: 2*Tt*: 1*tt*, or 1:2:1. The probability that an offspring will inherit the homozygous dominant genotype is 1 in 4. The offspring has a 2 in 4 (or 50%) chance of inheriting the heterozygous genotype *Tt*, and a 1 in 4 chance of inheriting the homozygous genotype of *tt*. In other words, any offspring of this cross has a 75% chance of being tall and a 25% chance of being short.

Female

Male	T	t
T	TT	Tt
t	Tt	tt

FURTHER EXPLORATION
A **dihybrid cross** involves more than one genotype. Work with a partner to find out more about dihybrid crosses. Then create a Punnett square of your own to illustrate a dihybrid cross.

Female

Male	TG	Tg	tG	tg
TG	TTGG	TTGg	TtGG	TtGg
Tg	TTGg	TTgg	TtGg	Ttgg
tG	TtGG	TtGg	ttGG	ttGg
tg	TtGg	Ttgg	ttGg	ttgg

T = Tall   t = short   G = Green   g = Yellow

ATOMS

*A*ll matter is made of tiny building blocks called atoms.
Atoms are so small that if each person alive today was the size
of an atom, the entire population of the world would fit
on the head of a pin.

What Are Atoms?
All matter is made of one or more basic substances called **chemical elements**. Oxygen, gold, carbon, and hydrogen are all elements. There are 109 known chemical elements either occurring in nature or manufactured in laboratories.

An **atom** is the smallest particle of an element having the unique chemical properties of that element. For instance, if you divide a nugget of gold in half, you will have two pieces of gold. If you continue to divide the nugget into smaller and smaller pieces, the smallest particle of gold that you can divide the nugget into and still have gold is an atom of gold. If you divide that atom, it will no longer have the properties of gold.

The ancient Greeks first conceived of the atom. The philosopher Democritus coined the name *atom* from the Greek word *atomos* which means,"that which cannot be split." For more than two thousand years, evidence suggested that atoms were hard solid particles that could not be divided into smaller particles. Starting in the early 1900's, physicists conceived the present-day model of the atom.

Subatomic Particles
Experimental evidence in the late 1800's and throughout the 1900's suggested to scientists that atoms consist of a tiny, dense, positively-charged core called the **nucleus** surrounded by a much larger, negatively-charged space called the **electron cloud**. Most of the volume of an atom consists of the electron cloud, which is mostly empty space. In fact, if the nucleus of an atom was the size of a ping-pong ball, the entire atom would be a mile in diameter.

The particles that make up the atom are known as subatomic particles. The three major **subatomic particles** are protons, neutrons, and electrons. The densely packed nucleus of the atom is made of protons and neutrons, and accounts for almost the entire mass of the atom. In fact, the nucleus is so dense that a small marble made of only protons and neutrons would weigh 100 million tons.

Electrons are negatively charged particles and are far less massive than protons and neutrons. **Protons** are positively charged particles that weigh about 1840 times as much as electrons. The proton's positive charge is equal in magnitude to the negative charge of the electron. **Neutrons** are slightly heavier than protons and have no overall electric charge, that is, they are neutral. All atoms, except for hydrogen, have neutrons. Hydrogen has only one proton and one electron, and is the simplest and most abundant element.

FUNTASTIC FACTS

▪ A process called Carbon-14 dating is used to determine the age of prehistoric organisms. Scientists compare the ratio of Carbon-12 isotopes to Carbon-14 isotopes present in the organism.

Isotopes

All atoms of the same element have the same number of protons and electrons, therefore, the net electric charge on an atom is zero. In fact, an element is defined by the number of protons in its atom's nucleus. For instance, all atoms of carbon have six protons, and all atoms of oxygen have eight protons. However, it is possible for the same element to have atoms with different numbers of neutrons. Atoms of the same element having different numbers of neutrons are called **isotopes** of that element. The most common isotope of carbon is Carbon-12 with six protons and six neutrons. The less common isotope Carbon-14 has six protons and eight neutrons.

Scientists have learned how to create isotopes in the laboratory that do not occur in nature. These isotopes, however, turn out to be very unstable. For example, the element copper has two natural isotopes, one containing 34 neutrons, and the other containing 36 neutrons. Scientists have bombarded copper atoms with high velocity neutrons creating a number of artificial isotopes, such as copper with 35 neutrons. This nucleus is unstable and lasts only about thirteen hours. A copper isotope with 33 neutrons lasts only about eleven minutes, and one with 37 neutrons lasts about five minutes. Evidently, nature has some very definite rules about the number of neutrons that should be in a stable nucleus.

Electron Energy Levels

Electrons within the electron cloud exist in one or more **energy levels,** or energy states. Each energy level can only be occupied by a specific number of electrons at any one time. The lowest energy level can be occupied by two electrons. The next two energy levels can be occupied by up to eight electrons. In general, electrons "fill up" one energy level before occupying the next highest level.

When an atom receives enough energy, its electrons can suddenly "jump" to higher energy levels. When this happens, the atom is said to be excited. As an excited atom loses energy, the electrons fall back to lower energy levels and emit a definite wavelength of energy corresponding to a distinct color of light. Excited atoms are responsible for the brilliant colors that we see in fireworks.

READ MORE ABOUT CHEMISTRY

● Asimov, I.
A Short History of Chemistry.
Garden City, New York: Doubleday and Co.,Inc., 1965.

● Berger, M.
Atoms, Molecules, and Quarks.
Toronto: General Publishing Company Limited, 1986.

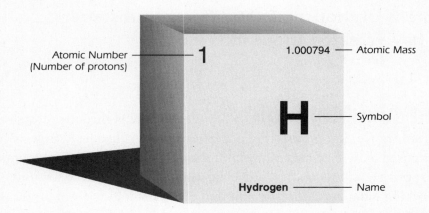

Atomic Number (Number of protons) — 1

1.000794 — Atomic Mass

H — Symbol

Hydrogen — Name

Hydrogen, the first of the 109 elements in the Periodic Table of the Elements

FURTHER EXPLORATION

The isotopes artificially created by scientists in the laboratory constitute a new and effective tool in the search for knowledge. These isotopes are extremely useful in medical research and medical treatments, such as radiation treatment for cancer patients. Work in groups to find out other uses for radioactive isotopes.

MOLECULES

*E*verything in our world exists because different atoms join together to form particles called molecules. From the basic elements, nature has produced hundreds of thousands of different substances; people have created thousands more.

What Are Molecules?

A **molecule** is the smallest particle of a substance that retains the properties of that substance. Molecules are composed of at least two atoms, but may contain thousands of atoms linked together. Atoms of the same element sometimes join to form molecules of that element. Atoms of different elements join to form molecules called **compounds**. For example, two atoms of hydrogen join together to form a molecule of hydrogen. Two atoms of hydrogen and one of oxygen join together to form the compound water.

What Holds Molecules Together?

The process of joining atoms together is called **bonding**. The two most common forms of bonding are ionic bonding and covalent bonding. In **ionic bonding**, there is a transfer of electrons from one atom to another. In **covalent bonding**, the atoms share electrons. The type of bond that occurs is usually the one that provides the most stability for the atoms involved.

Electrons surround the nucleus of an atom at definite energy levels called **electron shells**. Complete electron shells are very stable structures, and atoms tend to combine in ways that complete their outer electron shells.

The Ionic Bond

Ionic bonding occurs in the formation of many substances, including salt. Ordinary table salt consists of molecules of sodium chloride. Atoms of sodium have one more electron than a complete outer electron shell. Atoms of chlorine are short one electron to complete an electron shell. Therefore, when sodium and chlorine join to

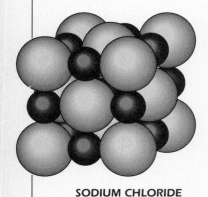

SODIUM CHLORIDE

form sodium chloride, a sodium atom donates an electron to a chlorine atom. This completes the electron shells of each atom and forms a stable molecule of sodium chloride.

You may wonder what actually holds the two atoms together. Atoms of sodium, like all other atoms, are electrically neutral. That is, they have an equal number of positively charged protons and negatively charged electrons, resulting in a net charge of zero. When an atom of sodium donates one electron, the sodium atom is left with a net positive charge of plus one. Similarly, when a chlorine atom receives an extra electron, it is left with a net negative charge of minus one. Atoms with net charges other than zero are called **ions**. Since opposite electrical charges attract, the positive sodium ion and the negative chlorine ion are pulled together.

The Covalent Bond
More common than the ionic bond is the covalent bond. A good example of a covalent bond occurring is with the simplest molecule, hydrogen. A molecule of hydrogen consists of two atoms of hydrogen that share electrons. Each hydrogen atom has one proton and one electron and needs one electron to complete its first electron shell. As two hydrogen atoms come together, their electrons are shared. This sharing completes the electron shells of both atoms. The atoms are held together by electrons occupying a molecular orbital.

HYDROGEN MOLECULE

Phases Of Substances
The three forms or **phases** in which a substance can exist are gas, liquid, and solid. The phase of a substance depends largely upon the distances between its molecules. In a gas, molecules are very far apart and travel freely. Under normal conditions, most of the volume occupied by a gas is empty space. The molecules of a solid are so close together that their movements are restricted to vibrating back and forth. Although molecules of a liquid are in close contact, they are able to slide over one another like a large group of marbles in a bag. It is this property that enables liquids to flow.

The Language Of Chemistry
In the nineteenth century, chemists began using letters and combinations of letters to symbolize the elements. Each lettered symbol represents one element. Using this system of chemical notation, it is possible to denote molecules and to describe chemical reactions using equations. For example, the symbol for a single atom of hydrogen is H, and the symbol for the two-atom

FUNTASTIC FACTS

- There are billions of Molecules in a single grain of salt.

- There are more molecules in a glass of water than there are grains of sand on all the beaches of the world.

molecule of hydrogen is H$_2$. To show how water (H$_2$O) is formed by combining molcules of hydrogen and oxygen, we write:

2H$_2$ + O$_2$ → 2H$_2$O. This expression means two molecules of hydrogen combine with one molecule of oxygen to form two molecules of water.

THE FORMATION OF WATER

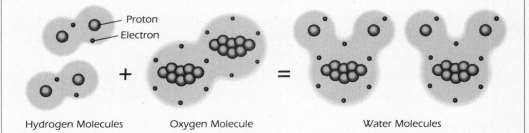

Hydrogen Molecules   Oxygen Molecule   Water Molecules

CAREERS IN CHEMISTRY

Chemists and chemical engineers help to develop new products and improve old ones. Most chemists do research for corporations that produce such things as pharmaceuticals, cosmetics, plastics, and foods. Chemical engineers often specialize in fields such as petroleum engineering and environmental science. Many chemists and chemical engineers also have backgrounds in other branches of science and engineering. Most chemists receive graduate or professional degrees, and many pursue Ph.D's.

FURTHER EXPLORATION

The two most common forms of molecular bonds are covalent bonds and ionic bonds. Work with a partner to find out what some other types of chemical bonds are and how they join atoms together.

RADIOACTIVE DECAY

*T*he study of radioactivity has led to advances in medicine, manufacturing, scientific research, and energy production. It has also resulted in the creation of the most dangerous devices on Earth—nuclear weapons.

What Is A Half-Life?

All chemical elements heavier than lead (atomic number 82) are unstable. They break down by emitting nuclear particles and losing energy. These unstable elements are called **radioactive**.

The breakdown of radioactive substances, such as uranium and plutonium, is **radioactive decay**. Scientists have discovered that different samples of the same radioactive material will always break down at the same rate. For example, after 24 days, 20 ounces of the isotope Uranium-239 will reduce to only 10 ounces of the same substance. After another 24 days, only 5 ounces will remain, and after an additional 24 days, only 2.5 ounces of the uranium will be left.

The 24-day period it takes for Uranium-239 to lose half its mass is known as its half-life. The **half-life** of an element is the time it takes for half the element's mass to change by radioactive decay. The half-life of some isotopes are as brief as a few millionths of a second. The half-life of other isotopes may be billions of years.

The Breakdown Of The Nucleus

When a radioactive element decays, changes occur in its nucleus. For example, radium decays naturally to form the more stable element radon. The radium atom nucleus undergoes change in which its energy decreases and it obtains a more stable form. This type of change is known as a **nuclear reaction**. A nuclear reaction differs from a chemical reaction. In a chemical reaction, changes take place in the arrangement of electrons, but there are no changes in nuclei. When the nucleus of a radium atom emits an alpha particle, radon is formed from the decaying radium atom. An **alpha particle** consists of two protons and two neutrons travelling at about 10,000 miles per second. An alpha particle is only one kind of radiation. Beta particles and gamma rays are other kinds.

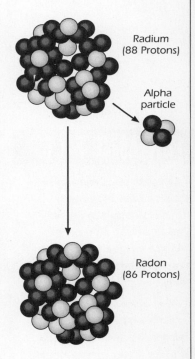

Radium
(88 Protons)

Alpha particle

Radon
(86 Protons)

COMPUTERS

*C*omputers play an ever-increasing role in today's information society. They perform many tasks thousands of times, sometimes millions of times faster than a human being.

FUNTASTIC FACTS

▪ **Compact discs contain bits which a laser reads and translates into the music that you hear. Millions of bits are needed to store an average-length four minute song.**

▪ **Fiber optic telephones transfer bits using a light beam and a thin wire made of glass. Many conventional phone systems in the United States are being replaced with better quality fiber optic systems.**

Bits

Computers operate on a basic principle of electricity. Currents can be either on or off. For example, when someone turns on a light switch, electricity flows to a light bulb, and the light goes on. When the same switch is turned off, the electricity stops flowing and the light goes off. Using the same switch, it is impossible to turn a light partially on.

A **bit** is an electrical storage element in a computer represented by either a one or a zero. 1 represents *on* and 0 represents *off*. A computer chip can only distinguish between two energy levels, *on* and *off*.

Bits in a computer are a little like cells in a human being. They are the building blocks of a computer's operations. Every function a computer performs is based on the concept of a bit. A computer intensively processes bits to perform addition, subtraction, and other calculations.

This system of bits, or ons and offs, or ones and zeros, is also referred to as the **binary number system**.

Bytes

Unfortunately, you do not perform calculations using the binary number system as computers do. You calculate using **decimal numbers**, numbers based on the number 10. So how does a computer represent a decimal number, such as 23, when it only uses 1's and 0's? It uses bytes. A **byte** (pronounced bite) is a memory location made up of eight bits. The unique combinations of these bits in a byte determines the binary system representation of a decimal number.

Counting In Binary

Counting in the binary system is based on the same principle of counting in the decimal system. When counting from 0 in the decimal system, you start at 0, count to 9, move to the next place value, and count up again from 0, and so on. When counting from 0 in the binary system, you start at 0, count to 1, move to the next place value, and count up again from 0.

The following table illustrates counting from 1 to 10 in both the decimal and binary number systems.

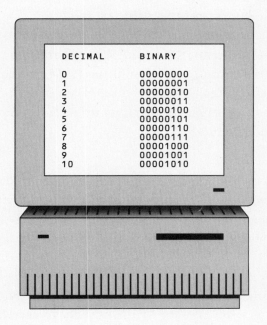

```
DECIMAL        BINARY

0              00000000
1              00000001
2              00000010
3              00000011
4              00000100
5              00000101
6              00000110
7              00000111
8              00001000
9              00001001
10             00001010
```

Each place value in the decimal system is based on 10. Each place value in the binary system is based on 2. The following place value chart shows the number 100 in decimal and binary.

DECIMAL

10^3	10^2	10	1
0	1	0	0

BINARY

2^7	2^6	2^5	2^4	2^3	2^2	2	1
0	1	1	0	0	1	0	0

Rom And Ram

Of course, computers have a way of taking decimal numbers you type on a computer keyboard and converting them to 1's and 0's so it can perform calculations on them.

All computers have memory made up of many bits. Information is stored in and later retrieved from a computer's memory. Computers have two kinds of memory: **ROM**, or Read Only Memory, and **RAM**, or Random Access Memory. ROM stays intact even when a computer is turned off. It is also referred to as **non-volatile memory**. Information in RAM can easily be changed or erased. RAM is sometimes called **volatile memory**.

When you type any key on a computer keyboard, the computer looks immediately in its Read Only Memory (ROM) and converts the number or letter you typed into a byte. The byte is then transferred into a Random Access Memory (RAM) location where it is processed further depending on whether you typed a letter or number. Letters, or characters, are simply stored in a memory location. The computer performs calculations on numbers before they are stored.

How does the computer transfer what you typed on the keyboard into a graphic image on the computer screen? After characters or numbers have been stored in a memory location in the RAM, the computer compares the new byte to bytes listed in the ROM. Information in the ROM is then used to place a number or character on your monitor by turning on or off dots called **pixels**.

Pixels
The computer monitor is essential for you to see what you have typed or to see what the computer has calculated. The display on a monitor is made up of thousands of small dots called **pixels**. On a monochrome monitor, where only one color can be displayed, a pixel can be either on or off.

Pixels are turned on and off in certain ways to display letters, numbers, and graphic images. The **resolution**, or visible sharpness of numbers, letters, and pictures is determined by the number of pixels a computer has. The most common number of pixels on a home computer is 640 horizontal pixels by 400 vertical pixels, or 256,000 pixels in total.

On a computer system with a monochrome monitor, the number of bits needed to store a picture on the screen equals the number of pixels on the screen. On most home computers, it would take 256,000 bits, or 32,000 bytes to store a picture.

Color Graphics
When something is displayed on a color computer monitor, more bits are required to store a picture than on a monochrome monitor.

For example: Suppose a computer can display 16 different colors and has 512,000 pixels. How many memory bits are required to store the picture?

Step 1: One bit is always needed to store each pixel.

Step 2: The 16 colors can be thought of as being numbered 0 through 15. The decimal number fifteen is represented by the binary numeral 1111, or four bits.

Step 3: Altogether, it takes 5 bits per pixel to store information on this computer. It takes 2,560,000 bits, or 320,000 bytes to store the picture.

The greater the number of pixels and colors a computer system has, the more bits are needed to display a picture.

FURTHER EXPLORATION

Discuss with a partner how you might multiply two binary numbers. Work out an example to illustrate how to do it.

FRACTALS

A **fractal** is often defined as a graphical representation of chaos. Fractals make computer graphics more versatile and realistic. By using fractals, mountain ranges can look craggier and more menacing to climb, while ocean waves may appear more choppy and unpredictable.

TORNADOS

More proof that weather is chaotic is the fact that tornados cannot be predicted even though meteorologists are aware of the conditions that cause them.

What Is Chaos?

Think about the weather. It is impossible for anyone to predict the weather 100% of the time. For that reason, weather may be considered chaotic. Of course, you could say that weather is not truly chaotic because tomorrow's weather can be predicted accurately most of the time. Unfortunately, this statement is not quite true when you consider how weather is forecast.

When you watch the local television news, a meteorologist predicts the weather for your particular area, which could range from five square miles or less to one thousand square miles or more. What are the odds that the weather prediction will hold true for your backyard? When you take those odds into consideration together with the fact that no meteorologist can predict rain at an exact day and time, you must conclude that weather patterns are in fact chaotic.

Mathematically speaking, **chaos** can be defined as something that does not have an answer found by using a simple formula. It is easy to calculate how much rain has fallen on an area, but it is impossible to calculate how much rain will fall before the rain has started.

Chaos And Fractals

Fractals are essentially pictures of chaos. In a way, this seems contradictory. Graphs usually represent equations or show statistical information. So how is it possible to graph something chaotic? If you took a piece of graph paper and decided to graph chaos, where would you start? Wouldn't the graph paper eventually look like a clutter of points and lines that mean nothing? At first glance, this may seem to be true. The following example, however, may give you a different perspective on graphing chaos.

Foxes And Rabbits

A classic mathematical example of chaos is the foxes and rabbits scenario. Suppose there is a forest where there are only rabbits, foxes, and vegetation. You are told that in January of this year, there were 100 rabbits and 89 foxes in the forest and that vegetation was plentiful. You are instructed to graph the rabbit and fox population for the next 10 years versus time on a piece of graph paper. If the person who requested this graph expects exact results, he or she is sadly misguided.

The rabbit population is related to the fox population and to the amount of vegetation. On the other hand, the number of foxes is related to the number of rabbits on which to prey. As you can see, there is no way to predict what the populations will be at any one point.

The Power Of The Computer

Fortunately, a computer program can graph these populations. Without the aid of a computer, this problem would prove to be extremely tedious.

The first thing such a program does is to read the initial fox and rabbit populations. Second, the program reads the number of years to graph. The rest of the program essentially involves randomization.

Why must the computer randomize? Something must occur to affect the population of either the foxes or rabbits. For example, suppose that a fox kills three rabbits, and there are 97 rabbits left. This reduction in rabbit population can affect the population of foxes since the foxes need the rabbits for food. After the fox population decreases because of this, the rabbit population will again begin to grow, and so on. Because there is randomness in the computer program, it is highly unlikely that the same results will occur in more than one run of the program. Through randomization, the computer creates hypothetical situations to show chaos. This graph shows the population scenario that one run of the computer program could produce.

READ MORE ABOUT COMPUTERS
- Borse, G.J. **Fortran 77 and Numerical Methods for Engineers**. Boston: Prindle, Weber, and Schimidt Publishers, 1985.
- Gleick, James. **Chaos, Making A New Science**. New York: Penguin Books, 1988.

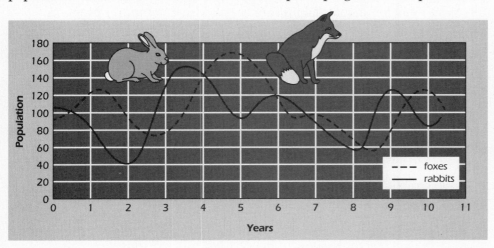

Applications

Examine the population graph. Other than serving as a picture of a hypothetical situation about fox and rabbit populations, what other uses might the shape of this graph have?

Suppose a programmer wished to create a computer game in which there are mountains. The programmer could simply use the graph of the fox population to look like a mountain range. Every time the game is played, the mountain formations could change. This is a simple example of how fractals can be used in creating computer graphics.

The Mandelbrot Set

One of the most famous fractals is the Mandelbrot Set, which was first presented by Benoit Mandelbrot. The following is an illustration of this fractal.

Practical Uses For Fractals

Fractals are mainly used to produce realistic and exciting computer graphics. They are often excellent imitations of things we see in nature.

When you see a detailed picture of a tree in a computer program or game, there is a good chance that it was originally generated using a fractal. Obviously, trees do not grow in an exact pattern. Their growth patterns are chaotic, and can be well represented by fractals. Of course, it is not possible to represent every type of chaotic behavior with a fractal. For example, biological systems in the human body behave unpredictably and chaotically in a number of respects. There is not necessarily a good way to represent these systems using fractals.

FURTHER EXPLORATION

The Koch Snowflake, which begins as an equilateral triangle, is another famous fractal. Work in groups to research and explore how the Koch Snowflake is formed.

ACID RAIN

*P*ollutants from automobiles, factories, power plants, and many other sources have damaged our environment, perhaps irreversibly. Acid rain is a subtle but very destructive form of this pollution.

What Is Acid Rain?

Acid rain forms when rain is polluted by sulfur and nitric acids. Polluting sources, such as factories, power plants, and automobiles, burn fossil fuels (coal, oil, and gasoline) that react with the water vapor in the air to form acids. Acid rain kills fish and other forms of aquatic life, and can severely damage crops, trees, and soil. It evens corrodes buildings and statues. Acidic gases and particles in our atmosphere also fall to Earth even when there is no liquid precipitation.

Effects Of Acid Rain On Trees

Acid rain and acid fog and clouds have had a devastating effect on trees in the northeastern United States and Canada, especially those in high elevations. Many trees are stunted in growth, have fungus growing on their bark, and have leaves that are turning orange and dropping too early in autumn.

In 1991, researchers who studied sugar maples on Camel's Hump peak in the Green Mountains of Vermont discovered a 25 percent decline in the average height of the trees and a decline of almost 50 percent in the number of new trees over the last twenty-five years. (*The New York Times*) Some scientists see this as the beginning of an epidemic and expect a region-wide decline in forests within the next 50 to 100 years.

How Research Is Done

By cutting out a cross section of a tree, scientists can examine the concentric circles that form as a tree grows. Each circle represents one year of growth. Researchers in Vermont have analyzed chemical traces in the rings of 200 year old trees. They found almost no trace of chemical pollutants in wood grown in the early 1800's. There were some chemical traces found in wood grown

THE pH SCALE

Chemists and other scientists measure acidity by using the pH scale. More specifically, the **pH scale** is used to describe the amount of hydrogen ions in a solution. The letters **pH** mean "potential of hydrogen." The scale ranges from 0 to 14. Any number below 7 indicates that the solution is an acid: the lower the number, the stronger the acid. Any number above 7 is a basic or **alkaline** solution: the higher the number, the stronger the alkaline. The number 7 indicates a **neutral** solution, such as pure water.

WORD MEANINGS

- **acid**
  acidus (LATIN)
  sour

- **alkali**
  al (ARABIC)
  the
  quliy (ARABIC)
  ashes of saltwort, a
  salt–marsh plant

later that century when industrial pollution began to increase. Samples of wood examined from the 1950's and later show evidence of much greater chemical pollution.

By analyzing chemicals in the air and comparing them with wind currents, researchers in Vermont have identified the Ohio River Valley in the Midwest and, to a lesser extent, urban centers on the East Coast as the sources of much of the air pollution and acid rain in northern New England.

Coal burning factories and power plants in the Midwest emit many chemicals that help create acid rain. Prevailing wind currents carry these chemicals over 500 miles to the forests, lakes, and streams of New England and New York State.

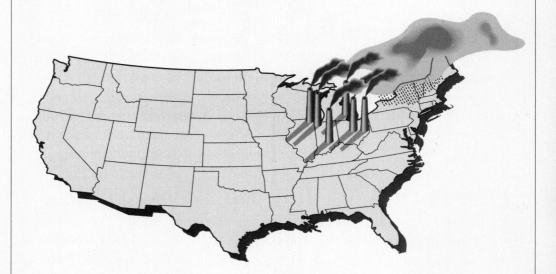

Reducing Acid Rain
We can reduce acid rain by taking a few steps. **1.** We can enforce current anti-pollution laws more strictly. **2.** We can install **scrubbers**, devices used to remove pollutants, in industrial smokestacks. **3.** We can search for ways to reduce, and eventually eliminate, the burning of coal because of its high sulfur content. To partially counter existing effects of acid rain, we can add lime to lakes, rivers, and soil to help neutralize acids.

FURTHER EXPLORATION
Acid rain is a national and international problem. Research and discuss the political difficulties in reducing the causes of acid rain.

OZONE DEPLETION

*H*igh in Earth's stratosphere, a layer of ozone blocks the sun's harmful, skin cancer-causing ultraviolet rays. In recent years, scientists have become alarmed that this protective ozone layer is being destroyed.

Ozone is a form of oxygen different from that which we usually breathe. Three oxygen atoms joined together form ozone. Two oxygen atoms joined together form breathable oxygen.

Even though ozone exists near Earth's surface, most ozone is found in the upper atmosphere. The highest concentrations of ozone are found between 9 and 18 miles above the surface of Earth. The **ozone layer** in the upper atmosphere makes life possible on Earth because it protects life from harmful rays of the sun. It blocks 95 percent to 99 percent of the sun's ultraviolet rays, the rays that are known to cause skin cancer.

Ultraviolet rays are dangerous to humans, animals, and plant life. In addition to higher levels of skin cancer, overexposure to ultraviolet rays can produce cataracts and severe damage to the immune system. Because of this danger, scientists constantly monitor ozone levels in the upper atmosphere.

A Hole In The Ozone Layer

Since the late 1970's, scientists have detected a hole in the ozone layer occurring above Antarctica every spring. The ozone there thins dramatically each spring, with over 40 percent of the ozone being depleted. As a result of the ozone hole discovery, CFC's, or **chlorofluorocarbons,** (chlorine-based chemicals primarily responsible for the damage to the ozone layer), were banned in aerosol spray cans made in the United States. However, many countries refused to ban CFC's, and leading manufacturers in the United States continued to sell CFC's for other uses, including blowing foams used in insulation, and as cushioning in running shoes.

CFC'S

Industries in the United States and the rest of the industrialized world began using CFC's, or chloroflourocarbons, in the 1930's. CFC's were used as propellants in hair spray and deodorant spray cans, and were employed in chemicals such as freon in refrigerators and air conditioners. Today, CFC's are being used less, and are no longer found in the contents of aerosol cans sold in the United States. Scientists are researching new chemicals to use in refrigerators and air conditioners. In the meantime, many developing countries, in which air conditioning and refrigeration are only now becoming widely used, have announced their opposition to the banning of CFC's.

Once CFC's are released, they rise into the atmosphere where the sun's ultraviolet light breaks them down into smaller parts. These parts react with ozone, reducing the ozone layer's concentration.

The Spy Plane Mission
In 1974, F. Sherwood Rowland and Mario Molina, two chemists at the University of California in Irvine, suggested that man-made chlorine compounds such as CFC's are being torn apart when they rise to the Earth's stratosphere by solar ultraviolet rays. Roland and Molina suspected that ultraviolet radiation frees pure chlorine and that chlorine attacks the natural layer of ozone that protects those elevations over Earth from ultraviolet rays.

In 1987, NASA decided to test Rowland's and Molina's theory. NASA sent 150 scientists and technicians to Punta Arenas at the southern tip of South America to determine possible causes for the destruction of the ozone layer. A modified U-2 spy plane capable of reaching altitudes exceeding 70,000 feet entered the ozone hole in the stratosphere. NASA discovered, with great apprehension, that the level of chlorine monoxide was 500 times more than is normal at Earth's mid-latitudes.

In 1991, the Environmental Protection Agency predicted that in the next fifty years we will see at least 12 million new cases of skin cancer and 200,000 more deaths resulting from skin cancer due mostly to the thinning of the ozone layer.

In January of 1992, the highest levels of chlorine monoxide ever recorded until that time were discovered over northern latitudes, appearing as far south as New England and eastern Canada (*The Los Angeles Times*). This research, conducted by NASA, the National Oceanic and Atmospheric Administration and the National Science Foundation, set off new alarms.

What We Can Do
In November, 1992, 74 nations were due to meet in Copenhagen. The purpose of the meeting was to update an international agreement to ban CFC's. During a previous meeting, the Montreal-Protocol was signed, setting a deadline for the banning of CFC's by the year 2000. It is anticipated that the deadline will be changed to 1995.

Global Warming
Many scientists believe that increased amounts of carbon dioxide and other chemicals in the air are bringing about a **greenhouse effect**, a trapping of the sun's heat by harmful gases

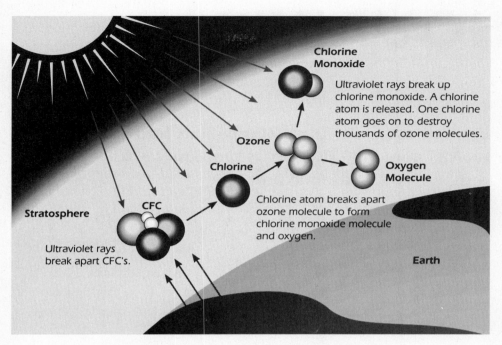

HOW THE OZONE LAYER IS BEING DESTROYED

Within the diagram:

Chlorine Monoxide — Ultraviolet rays break up chlorine monoxide. A chlorine atom is released. One chlorine atom goes on to destroy thousands of ozone molecules.

Ozone

Chlorine

Oxygen Molecule — Chlorine atom breaks apart ozone molecule to form chlorine monoxide molecule and oxygen.

Stratosphere

CFC — Ultraviolet rays break apart CFC's.

Earth

FUNTASTIC FACTS

■ In the late 1970's, CFC's were banned in the U. S. for use in aerosol cans. In 1992, chemical manufacturers such as DuPont were still selling CFC's for use in insulation, foam for running shoes, cleaner for electric circuit boards, and freon for car air conditioners. (LOS ANGELES TIMES MAGAZINE)

produced by industries and automobiles. Some scientists predict that the greenhouse effect will eventually cause Earth to heat irreversibly and make temperatures rise above levels that human beings can bear.

This trend in heightened global temperatures as a result of the greenhouse effect is called **global warming**. Although some observers feel that global warming is an immediate threat to the environment, others feel that there is little danger. Skeptics further argue that systematic weather data has only been in existence since 1880, and that it is nearly impossible to predict climate changes over the course of the next century.

However, computer-generated models predict that a likely doubling of carbon dioxide in the atmosphere during the next century will increase average temperatures on the planet by 3° to 9° Celsius. Detractors point to the fact that computers can only approximate climate conditions, and therefore may be inaccurate.

FURTHER EXPLORATION

Discuss what **biodiversity** means and explain why it would be a topic for discussion at international Earth summits.

EARTHQUAKES

Some earthquakes are as powerful as 200 million tons of TNT, or in other words, about 10,000 times more powerful than the first atomic bomb.

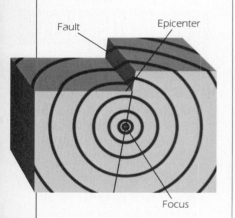

Fault   Epicenter

Focus

What Is An Earthquake?

An **earthquake** is a sudden shake or tremor in Earth's crust. As many as one million earthquakes may occur on Earth each year, most of them beneath the ocean. A large earthquake can be extremely powerful and destructive.

Earthquakes occur because of movements in Earth's crust. The outer crust of Earth is made up of **plates**, which are rigid, thick blocks. Earth's crust contains seven large plates and about 18 smaller ones. Rocks along plate boundaries are constantly squeezed and stretched. When the pressure and force on them become too great, the rocks break and shift, releasing energy in the form of an earthquake. **Faults** are the places where these ruptures occur. Most faults exist beneath the surface of Earth. Some, like the San Andreas Fault in California, can be seen on the surface.

The specific site of a sudden rock movement beneath Earth's surface is called the **focus**. Vibrations caused by earthquakes, called **seismic waves**, are strongest at the **epicenter**, the point on the surface directly above the focus.

The Richter Scale

A **seismograph** is an instrument that records Earth's ground motion. Scientists use the information obtained from a seismograph to calculate the magnitude of an earthquake on a scale of numbers called the **Richter scale**. Charles F. Richter developed this scale in 1935.

The highest magnitude ever recorded on the Richter scale was 8.9. About 1,000 earthquakes with a magnitude of 2.0 are recorded every day. Earthquakes with a magnitude of 5.0 or less are considered minor because they cause little or no damage. However, earthquakes with a magnitude of 7 or greater often cause major damage. The San Francisco earthquake of

1906 measured 8.3 on the Richter scale. Because of the loss of human life and devastation of property that resulted from it, the San Francisco quake is usually thought of as one of the worst earthquakes in history.

Recent California Earthquakes

Some California **seismologists**, people who study earthquakes, have grown concerned in recent years about a series of moderate quakes in the San Gabriel Valley east of Los Angeles. (*Science News*) These quakes have ranged in magnitude from 4.5 to 5.9, and have attracted attention because that region has not experienced quakes of this size for about fifty years.

Since 1987, scientists have observed a northern pattern to these quakes heading toward the San Andreas and Sierra Madre faults. These faults have the potential of unleashing earthquakes with a magnitude of 7.1, like the Loma Prieta quake that crushed parts of San Francisco in 1989.

Scientists are intrigued by the fact that four of these moderate quakes started at an unusually deep level of 12 kilometers below ground surface.

Possible Causes Of The Quakes

A possible explanation for this quake activity is that Earth's lower crust and mantle are sliding under the upper crust in a gradual northward movement. Seismologists involved in the study believe that a patch in the south slipped northward and then crept forward slowly, causing a ripple effect. This motion had enough stored energy to trigger a moderate earthquake.

For the 12 million people living in the Los Angeles region that might be affected by a large quake, researchers refuse to make specific predictions of how the seismic activity will end. The devastating 1989 Loma Prieta tremor began as a moderate quake. Other moderate earthquakes, however, have fizzled out. All researchers will say for sure is that people living in the region of potential danger should expect to feel more of the same kind of tremors they have experienced for the past few years. Whether or not these moderate tremors culminate in a major earthquake remains to be seen.

FUNTASTIC FACTS

- People can sometimes cause earthquakes. In the 1960's, tremors occurred in Denver, Colorado after people pumped large amounts of liquid waste into wells. The liquid seeped into faults, causing rocks to slip more easily.

- Earthquakes shake and loosen bricks in buildings and walls, causing these structures to collapse. Quakes may also damage water pipes, gas mains, and electric lines. The single-most dangerous result of an earthquake, however, is fire.

FURTHER EXPLORATION

Some earthquakes do not take place on plate boundaries. Prepare an oral report about the New Madrid fault in Missouri that caused major earthquakes in the early 1800's.

LIGHT

*T*he ancient Greek philosopher, Plato, theorized that light travels from a person's eye to an object. Pythagoras thought that light was made of particles that come from viewed objects. Today, exactly what light is still eludes scientists. It is possible, however, to observe and understand how light behaves.

FUNTASTIC FACTS

▪ **Light travels about 6,000,000,000,000 miles in a year. This distance is called a light-year.**

▪ **Some stars in the Milky Way Galaxy are so far away that the light waves they emit take many years to reach Earth. Light from other galaxies takes even longer.**

Light Waves

Light travels in waves. To understand this better, try the following experiment:

Hold one end of a rope and have a partner hold the other end. Pull the rope tight, leaving only a little slack. Have your partner abruptly yank her end of the rope up, then down. You will see that a hump travels from one end of the rope to the other. This movement in the rope is one kind of travelling wave.

The wave travelling through the rope is one model of a light wave. Light and other electromagnetic waves, however, originate at a source and travel through free space. An **electromagnetic wave** is a wave that is produced when an electric force accelerates charged atomic particles.

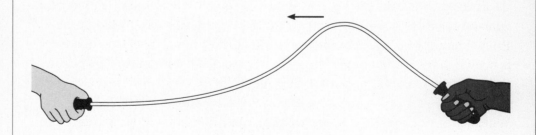

The Visible Spectrum

The length of a wave, measured from one valley to the next, is simply called its **wavelength**. The length of an electromagnetic wave determines where the wave falls in the **electromagnetic spectrum**. One kind of electromagnetic wave, X-rays, has exceptionally small wavelengths and cannot be seen. Light, which has wavelengths between 0.0000004 and 0.0000007 meters, falls into what is known as the **visible spectrum**.

Light of different wavelengths has different colors. Light on the low end of the visible spectrum, with the smallest wavelengths, is either violet or blue. Light on the high end of the spectrum, with the greatest wavelengths, is red or orange. Light that falls in the middle of the visible spectrum is either green or yellow.

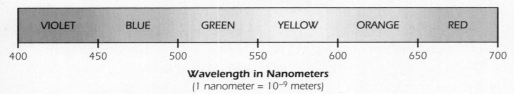

THE VISIBLE SPECTRUM

| VIOLET | BLUE | GREEN | YELLOW | ORANGE | RED |

400   450   500   550   600   650   700

Wavelength in Nanometers
(1 nanometer = 10^{-9} meters)

Polarized Light
The shape of the travelling wave in the rope experiment you performed was vertical. Light, however, does not travel in only one plane. In other words, light waves do not only travel vertically or horizontally. They can travel in many planes, (or in many different directions).

Polarized light is light that has some of its wave planes blocked. Many kinds of tinted automobile windshields polarize, or partially block, light waves from the sun by using polarizing films. Some films are created to block out more light than other films. More darkly tinted polarized windshields block more wave patterns than lightly tinted ones do.

The Speed of Light
Light travels faster than any object on Earth. The speed of light is about 186,000 miles per second (or about 300,000,000 meters per second). This speed is constant in a vacuum.

Galileo was the first scientist known to try measuring the speed of light. His assistant stood on a hill about one kilometer away. The assistant flashed a lantern as quickly as he could after he saw Galileo flash a lantern. Galileo timed the interval between the flashes of the two lanterns. The experimenters could not react fast enough, however, and Galileo had to conclude that light could not be measured using this method.

Light Reflection
Light does not always travel in a straight line. Often, surfaces it hits affect its path of travel. Light passes through **transparent** objects, such as glass. Light bounces off **opaque** surfaces, such as walls, buildings, trees, and people.

If light did not have the ability to **reflect**, or bounce off, surfaces, the world would be a much darker place. Every time you turn on a light switch, light emanates from the light source and then reflects off the lamp shade, the walls, and many other objects in the room. Little of the brightness you see in a room comes from the light bulb itself. Most of the brightness is actually a reflection of the light from various surfaces.

Some objects reflect light better than others. For example, a white wall reflects light much better than a black wall. Sand reflects light better than a piece of dark slate rock.

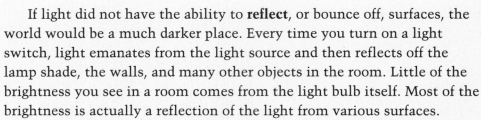

The Laws Of Reflection
Light is usually represented on paper by a ray, or arrow. The light ray that hits a reflecting surface, such as a plane mirror, is known as the **incident ray**. The **point of incidence** is where a ray hits the surface. The **normal** is a line constructed at right angles to the surface at the point of incidence. The light ray that is reflected by a surface is called the **reflected ray**. The angle between the incident ray and the normal (called the **angle of incidence**) is equal to the angle between the reflected ray and the normal (called the **angle of reflection**).

The **laws of reflection** when using a plane mirror are:
- The angle of incidence is equal to the angle of reflection.
- The incident ray, the reflected ray, and the normal all lie in the same plane.

You may wonder what happens when a light ray strikes a reflecting surface on the normal. The light reflects in the same path it travelled to the surface. Think of how bright light is when it is reflected from a mirror directly into your eyes.

Incident Angle = Reflected Angle

Light Refraction
Try this experiment: Place a penny in a clear glass of water. Stand about one foot away from the glass and notice how the penny appears to float. This effect is caused by **light refraction**, a change in direction of light as it travels at an angle from one substance to another.

Not all substances refract light to the same degree. Gelatin, for example, refracts light more than water. The speed of light is actually slower in an optically dense medium, such as gelatin, than it is through a medium, such as water or air, that is not as optically dense.

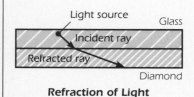

Refraction of Light

The Index Of Refraction

The amount of refraction that occurs between two given substances, such as air and water, or diamond and water, can be determined by using the index of refraction and by experimentation. The **index of refraction** (n) is a ratio of the speed of light in a vacuum (c) to the speed of light in a given material (v).

$$n = \frac{c}{v}$$

There is no way to determine the index of refraction without actually experimenting on the substances. The following table lists the index of refraction for various substances with respect to a vacuum. In other words, assume that light travels in a vacuum before it hits the refracting substance. The index of refraction is different when light travels through a substance and then encounters a vacuum.

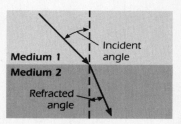

$$\frac{\text{Sine (incident angle)}}{\text{Sine (refracted angle)}} = \text{Index Of Refraction}$$

Substance	Index of Refraction
Air	1.00029
Sugar Solution (30%)	1.38000
Sugar Solution (80%)	1.49000
Salt	1.54000
Diamond	2.42000

Snell's Law

In 1621, the Dutch mathematician Willebrod Snell determined the exact relationship between the angle of incidence and the angle of refraction. **Snell's Law** states that the ratio of the sine of the angle of incidence to the sine of the angle of refraction is a constant for any substance. That constant is the index of refraction for that substance.

$$\frac{\sin i}{\sin R} = n$$

Since the incident angle and the index of refraction are known, the angle of the refracted ray can be determined.

FURTHER EXPLORATION

Work with a partner to investigate how lenses reflect and refract light. Discuss how concave and convex lenses differ. If possible, ask an optometrist about the uses of concave and convex lenses, and about the different indexes of refraction for glass and plastic.

NUTRITION

A nutritional diet not only ensures good health; it also provides you with shiny hair, clear skin, healthier teeth and gums, and a general feeling of well-being.

WORD MEANINGS
- **nutrient**
  nutrire (LATIN)
  to nourish
- **calorie**
  calor (LATIN)
  heat

What Is Nutrition?

The process by which plants, animals, and human beings use food is called **nutrition**. **Nutrients** are chemicals that supply an organism with the energy and substances necessary for growth. Carbohydrates, fats, and proteins are energy-supplying nutrients. Water, vitamins, and minerals are nutrients that are essential for survival, but do not supply energy.

Calories

A **calorie** is the amount of energy needed to raise the temperature of one gram of water by 1°Celsius. The energy in food is actually measured in kilocalories. One **food calorie** is equal to 1000 calories, or 1 kilocalorie.

Kinds Of Nutrients

Carbohydrates are starches and sugars, and are found in foods such as bread, cake, pasta, vegetables, and certain fruits. Carbohydrates provide about 60 percent of the calories in the diets of most Americans. Many nutritionists advise eating more complex carbohydrates, such as potatoes and pasta, and fewer simple carbohydrates, foods high in sugar.

Fats contain a great deal of energy in a small volume and are stored in your body for when food is not available. Between 30 percent and 40 percent of the calories Americans consume comes from fats. Foods from animals and from some plants, such as coconuts, provide most of the fats in the American diet. Many nutritionists suggest diets that are much lower in fats than the usual American diet.

Proteins are essential for growth and cell repair. Between 10 percent and 15 percent of the caloric intake of most Americans comes from proteins. Meat, poultry, fish, and many foods from plants, such as peanuts, are rich in proteins. Some nutritionists suggest that it is more healthful to get protein from beans and vegetables than from animal sources.

Minerals are inorganic substances necessary for growth and maintenance of your body. You can obtain most of the minerals you need from animal and plant foods, especially green vegetables and fruits. **Vitamins** come from all kinds of plant and animal food sources and are necessary for various chemical reactions to occur in the body. Water is part of what we are and of most things that we consume. Water has many functions in the body. One important function is to dissolve food so that it can pass into the bloodstream. Without water, human beings could not survive very long.

Gaining And Losing Weight
You gain or lose weight by eating more or fewer calories than you need and by decreasing or increasing activity. If you consume more calories than your body can use, the calories are converted to fat and stored in your body.

The number of calories that you need to maintain your weight depends on factors such as body size and composition, activity level, and age. Everyone uses calories at a different rate.

Doing certain activities will help you use calories more quickly than doing other activities. Running for one minute will help you use the calories of the apple you just ate much faster than by watching television for a half hour.

The following table lists some foods and the approximate number of minutes it would take one particular person to use the calories from these foods in a variety of ways.

NUMBER OF MINUTES NEEDED TO USE FOOD CALORIES

Food	Resting	Walking 2 1/2 mph	Walking 4 mph	Bike Riding 9 mph	Swimming	Running
Apple 125 cals	83	34	23	20	14	14
Fried Egg 95 cals	63	26	17	15	10	10
Milkshake 335 cals	223	91	61	54	37	37
1 Slice of Pizza 290 cals	193	79	53	46	32	32
Sirloin Steak (3 oz.) 240 cals	160	65	44	38	26	26
Strawberry Shortcake 417 cals	278	114	76	67	45	45

Activity

FURTHER EXPLORATION

Locate a reference on recommended calorie intakes. Based on your own height, weight, age, sex, and activity level, create a weekly menu that will enable you to maintain your weight while consuming foods that are nutritionally well-balanced.

GRAVITATION

*I*f gravitation were suddenly turned off, objects on Earth
would float into space, planets would leave their orbits,
and the universe would come apart.

FUNTASTIC FACTS

▪ **Someone who can jump
6 feet high on Earth can
jump...**

**2.5 feet high on Jupiter.
16 feet high on Mars.
36 feet high on the moon.**

What Is Gravitation?
In the seventeenth century, Sir Isaac Newton, the great English scientist and mathematician, stated an accurate scientific law of universal gravitation. He realized that the same force that makes an apple fall to Earth keeps the planets in their orbits. **Gravitation** can be loosely defined as the attraction that two bodies of matter have for each other. The word, **gravity**, is used more specifically to describe the attraction of Earth or another heavenly body for smaller objects.

Every planet and moon exerts a gravitational force on objects. If you have ever seen films of astronauts walking on the moon, then you know how effortlessly they seemed to jump high into the air. By watching those films you can see that gravitational force exists on the moon. If there were no gravity on the moon, a jumping astronaut would propel himself into space indefinitely.

Newton's Law Of Gravitation
Newton stated that all objects attract each other. The force of this attraction, according to Newton, depends on: **1.** how great the mass of each object is and **2.** the distance between the two objects. Attraction increases in proportion to the masses of the objects. Attraction decreases in proportion to the square of the distance between the centers of the two masses.

The law means that an object of greater mass exerts a stronger gravitational force on other objects than an object of lesser mass does. Also, an object that is closer exerts a stronger gravitational force than an object that is farther away. On Jupiter, for example, gravity is stronger than it is on Earth because Jupiter's mass is much greater than Earth's. Jupiter's gravitational attraction of our moon, however, is weaker than the Earth's attraction because the moon is much closer to Earth.

Newton's Second Law Of Motion

It may seem that the force of gravity is the same on every object. This is not the case. The force that gravity exerts on any body is directly related to the mass of the body. **Newton's Second Law** states that force is equal to the mass of an object multiplied by its acceleration.

$$F = ma$$

F is the force. *m* is the mass. *a* is acceleration.

If acceleration is due to gravity, then you can find the force that gravity exerts on a particular mass. For example, when a roller coaster car accelerates down a hill, its acceleration is due to gravity.

Acceleration

Acceleration due to gravity on Earth, often referred to as *g*, is 9.81 meters per second squared. This means that for every second an object is falling in a vacuum without air resistance, the speed of the falling object will increase by 9.81 meters per second. The following diagram illustrates the speeds a ball in free fall will achieve at one second intervals.

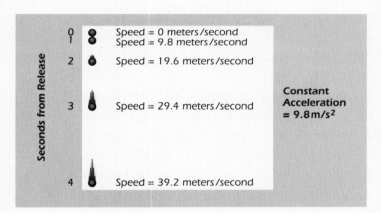

If the mass of an object is known, the gravitational force on that object can be determined by using *g*. By changing Newton's Second law, and using *g* as the constant for acceleration, you can determine the gravitational force on a given object.

$$F = mg$$

FURTHER EXPLORATION

Earth exerts a gravitational force on the moon, and the moon exerts a gravitational force on Earth. Discuss why the two bodies do not accelerate towards each other and collide.

FREE FALL

*I*f you hovered above the surface of the moon in a spacecraft and at the same moment dropped a hammer and a feather, both objects would fall with the same speed and acceleration and strike the ground at exactly the same instant.

From Aristotle To Galileo

Free fall describes the motion of an object falling towards the ground with or without air resistance. Modern scientific equations of motion do not include air resistance to predict the behavior of objects in free fall. Air, however, is almost always present on Earth and its effects on falling objects are not always obvious.

The ancient Greeks formed incorrect theories of motion because they did not consider the effects of air resistance. Aristotle believed that heavier objects fall faster than lighter ones. Until the sixteenth century, most scientists agreed with him.

Galileo Galilei, an Italian physicist and astronomer, was the first to question the ancient theories. Born in 1564, Galileo is considered by many to be the father of modern science. He conducted the first scientific experiments on freely falling bodies and distinguished between constant speeds and accelerations. He stated that bodies of different weights would fall equal distances in the same amount of time if the effects of air were neglected.

Galileo's Experiment

According to legend, Galileo dropped two objects of different weights from the top of the Tower of Pisa to observe the behavior of falling objects. In reality, it wasn't Galileo who used this method of observation. He didn't find it very useful. Scientists of the time using this method were at a great disadvantage because of the high speeds that the objects would achieve and the lack of instruments capable of accurately measuring their motion.

Galileo devised a new approach to studying the problem by rolling

FUNTASTIC FACTS

▪ **Objects on Earth fall about five times faster than they do on the moon.**

▪ **Without air resistance, airplanes could not fly and baseball pitchers could not throw curve balls.**

balls down inclined slopes. He realized that free fall was merely an extreme case of this motion in which the angle of the slope is increased to 90°. Galileo was able to slow the speeds over vertical distances enough to measure them accurately. In doing so, he was the first to create a mathematical model of acceleration. Galileo concluded that the vertical distance travelled increases in proportion to the square of time.

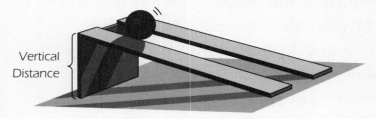

Vertical Distance

From Galileo To The Moon
About 400 years ago, Galileo predicted that a hammer and a feather dropped in the absence of air, from the same elevation, would strike the ground at exactly the same instant. His theory was tested during one of the NASA Apollo missions when a hammer and a feather were dropped to the surface of the moon. After four centuries, Galileo's ideas were proven correct.

CAREERS IN PHYSICS

Many physicists do research for universities, government agencies, and private corporations. Some of these scientists specialize in one or more branches of physics. Nuclear physicists study the atom and try to harness its power. Forensic physicists investigate the causes of accidents and disasters. Some physicists with strong backgrounds in computer science work as computer systems engineers and as corporate managers.

College freshmen wishing to major in physics should be well rounded students with a good grounding in mathematics and the sciences.

FURTHER EXPLORATION
Prepare an oral report on what effect air has on falling bodies. Find out what the terms **drag** and **terminal velocity** mean.

READ MORE ABOUT PHYSICS

- Halliday, D. and Resnick, R. **Fundamentals of Physics**. New York: John Wiley and Sons, 1981.

- McGrath, S. **Fun with Physics**. Washington ,D.C.: National Geographic Society, 1986.

- Thurber, W. **Exploring Physical Science**. Boston: Allyn and Bacon, 1977.

THROWN OBJECTS

*B*aseball, football, and basketball players always use the fundamental laws of physics to compete. Their brains instantaneously compute the quadratic equations necessary to bat a 100 mile per hour baseball, to throw a football at just the right speed, or to shoot a basket at precisely the correct angle.

Trajectory And The Parabola

The path that a thrown object takes is called its **trajectory**. The object is called a **projectile**. The trajectory of a thrown object follows the shape of the mathematical curve known as a parabola. To discover why the trajectory of a thrown object takes the shape of a parabola, you must first look at a few fundamental laws of physics.

Newton's Laws Of Physics

In 1687, Sir Isaac Newton, an English mathematician and scientist, stated that a body will continue in a state of rest or uniform motion (constant speed) in a straight line, unless acted upon by an external force. This statement is generally known as **Newton's First Law of Physics**. That means if you throw a ball into the air, it should travel in a straight line. You know, however, that its trajectory is that of a parabola. So an external force must be acting upon it. That force is gravity.

Why does the trajectory of a thrown object take the shape of a parabola instead of a semi-circle or a triangle? **Newton's Second Law of Physics** states that a force applied to a body causes it to accelerate in the direction in which the force is applied. Everyone knows that gravity works downward. If you jump up, you come down. As you throw the ball, gravity acts to slow, or decelerate, the upward speed of the ball until it stops rising, and then gravity accelerates the ball back to Earth. No force acts to slow the ball in the horizontal direction. According to Newton's First Law, the speed in the horizontal direction must remain constant.

WORD MEANINGS

- **trajectory**
  trans (LATIN)
  across

- **projectile**
  pro (LATIN)
  forward
  ject (LATIN)
  throw

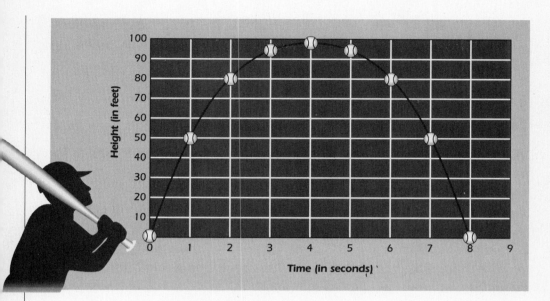

After looking at the diagram above, it is easy to see why the trajectory is a parabola. If each view of the ball represents one second of elapsed time, you can see that, for each second: 1) the ball travelled equal distances in the horizontal direction, representing constant speed, and 2) the ball travelled unequal distances in the vertical direction, representing acceleration. Therefore the path taken is a parabola.

Ballistics
Although trajectories follow a parabolic path, not all parabolas are the same. The parameters that determine the shape of a parabola are the initial speed of the projectile and the angle at which the projectile is launched. Knowing where a projectile is going to land can be extremely important, particularly to an army in wartime. The study of projectiles and the prediction of trajectories is known as **ballistics**.

The first systematic study of ballistics was conducted by the Italian scientist Niccolo Tartaglia in 1537. As a professional military engineer, he served as a consultant to several rulers and was commissioned to determine the angular elevation of a gun barrel that would yield the greatest range for a shot. Tartaglia correctly determined that a 45° angle of inclination was required.

Although the study of ballistics continued to develop after Tartaglia, the principles of this science could not be applied in practice because no acceptable method of measuring the initial velocity of a projectile existed. It was not until 1740 that Englishman Benjamin Robins invented the ballistic pendulum, a device capable of accurately measuring the velocity of

FUNTASTIC FACTS

▪ The dimples in a golf ball affect the airflow around the ball, reducing the air resistance and helping the ball to travel farther.

▪ If you alter the trajectory of a ball such that its path is no longer a parabola, it will be nearly impossible to catch. (This can be approximated by attaching an elastic band to a tennis ball.)

a projectile from a firearm. Over two hundred years after Tartaglia first studied projectile motion, ballistic theory finally began to make an impact on gun design. Today, the principles of ballistic theory are used to program the trajectories of guided rockets through space.

Outfielder Or Physicist?
An outfielder uses the principles of ballistics every time she performs the scientifically complex act of catching a fly ball. At the crack of the bat, using her sight and hearing, she estimates with amazing accuracy the initial velocity of the ball and the angle at which the ball leaves the bat. She instantaneously computes the trajectory of the ball and begins to run. All this happens in a fraction of a second. The outfielder is even able to adjust her calculations to compensate for such things as wind and air resistance. In a sense, we are all physicists.

FURTHER EXPLORATION
Prepare an oral report describing how trajectories and parabolas are used in everyday activities and in different professions.

PENDULUMS

*T*he most accurate clock in existence is the atomic clock, which is synchronized to the vibration of caesium atoms. It pulsates exactly 9,192,631,770 times per second.

The Quest For Accuracy

The ancestors of today's atomic clock can be traced back to pendulum, or grandfather, clocks. A **pendulum** is any swinging body, such as a chandelier, supported at a point. In fact, it was a swinging chandelier that led to the development of the pendulum clock.

While attending Mass in the cathedral at Pisa, the great mathematician and scientist, Galileo Galilei (1564-1642), observed a chandelier that was swinging slowly back and forth. He noticed that the time required for the fixture to complete one full swing was the same regardless of how large the swing was. Since he did not have a timepiece, he ingeniously timed the length of each swing by counting his pulse. Galileo's observations inspired him to study swinging bodies more closely.

The Simple Pendulum

A **simple pendulum** consists of a heavy weight and a string or relatively light rod. Galileo determined that the period of a pendulum depends only upon the length of the string or rod. The **period** of a pendulum refers to the time a pendulum takes to swing back and forth once. The **beat** of a pendulum refers to one half the period, or the time the pendulum takes to swing from its highest position on one side to its highest position on the other side. The beat of a pendulum one meter long is almost exactly one second. This means that a simple pendulum one meter long will have a period of two seconds regardless of how heavy the weight is. From these discoveries, Galileo reasoned that the pendulum could be used to measure time.

WORD MEANINGS
- **pendulum**
  pendulus (LATIN)
  hanging
- **period**
  periodos (GREEK)
  going around, cycle

Mechanical Clocks

Mechanical clocks are comprised of a system of gears driven by falling weights or springs. They were first developed in the thirteenth century. The accuracy of these clocks depends upon controlling the speed of the falling weights. An **escapement mechanism** does this by controlling the rotation of one gear, allowing the gear to move just one tooth at a time. The early clocks were so inaccurate that only a single hour hand appeared on the clock face. By the middle of the fifteenth century, clocks improved in accuracy, and a minute hand was used on some clock faces. A hand telling the seconds remained virtually unknown until the pendulum was used to measure time.

ESCAPEMENT MECHANISM

The First Pendulum Clock

Influenced by Galileo's ideas, Dutch scientist Christian Huygens invented the first swinging pendulum clock in 1657. His escapement mechanism consisted of an anchor attached to a pendulum that swung back and forth. As the anchor lifted on one side, a spring pushed the gear through until the next tooth caught on the anchor. As the pendulum continued its motion, the anchor pushed slightly backwards on the gear. The spring attached to the gear forced the anchor back in the other direction, maintaining the oscillations of the pendulum. Huygens's inventions are still used in clocks today.

Other Pendulums

Other examples of pendulums include torsional pendulums, conical pendulums, and compound pendulums. Instead of swinging back and forth, the weight of a **torsional pendulum** oscillates by twisting a rod back and forth. Many high precision laboratory instruments involve torsional oscillations, and some watches use tiny torsional pendulums to maintain accuracy.

TORSIONAL PENDULUM

A **conical pendulum** is similar to a simple pendulum because it consists of a weight attached to a string. The weight, however, travels around in a circle, forming a cone shape with the string. The period of the conical pendulum depends not only on the length of the string, but also on the angle of the string. Many amusement park rides are actually types of conical pendulums.

Get The Good Wood On The Ball

Have you ever wondered why a baseball bat sometimes stings a batter's hands as he hits the ball? Almost all objects that are swung, such as baseball bats, axes, and hammers, form compound pendulums. In simple pendulums, the suspending element, or string, is relatively light. A **compound**

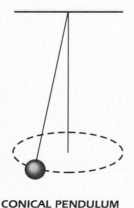

CONICAL PENDULUM

pendulum is one in which the weight of the suspending element, or handle, cannot be neglected.

If you swing a baseball bat from its handle and a simple pendulum of the same length, then you you will notice that the bat swings faster than the simple pendulum. This happens because the weight of the compound pendulum is distributed along its entire length instead of concentrated at its end, as in a simple pendulum. If you shorten the simple pendulum, you can find a length that produces a rate of swing equal to the rate of swing of the baseball bat. If you measure this length on the baseball bat from its handle, you will locate what physicists call the center of percussion. The **center of percussion** is where the weight of the bat may be considered concentrated for study as a simple pendulum. When a coach says, "get the good wood on the ball," what he or she really means is to strike the ball with the center of percussion of the bat. When the ball strikes at some point other than the center of percussion, the bat will shake and vibrate, causing the batter's hands to sting.

FUNTASTIC FACTS

- The first known clock was the sundial used by the ancient Egyptians more than 3000 years ago.

- The atomic clock is more consistent than the rotation of Earth.

FURTHER EXPLORATION

In 1851, French physicist Jean Foucault, using a freely suspended pendulum, proved that Earth rotates about its axis. He demonstrated his experiments to the public at the Pantheon in Paris. You may have seen a Foucault Pendulum at the Smithsonian Institute in Washington, D.C.

Work with a partner. Find out what a Foucault Pendulum is and how Jean Foucault used it to prove that Earth rotates on its axis.

ENERGY

*E*nergy is involved when a football is thrown, when a weight is lifted, when a spring is compressed or stretched, when an electron moves through an electric circuit, or when a bomb explodes.

Maximum Potential Energy
No Kinetic Energy

No Potential Energy
Maximum Kinetic Energy

What Is Energy?

Energy can be defined as anything that is not matter and that can cause a change in matter. **Kinetic energy** is the energy an object has because of its motion. **Potential energy** is the energy that an object has because of its position. For example, a rock at the top of a cliff has potential energy because the rock has the *potential* of releasing energy as it falls to the bottom of the cliff. **Work** is done on an object when a force changes the kinetic or potential energy of the object. When you lift a heavy object, the object gains potential energy: you are doing work on the object. If you release the object, the energy is converted from potential to kinetic energy as the object falls.

As the object falls towards the ground and its speed increases, the object loses potential energy and gains kinetic energy. Just before hitting the ground, the object possesses its maximum amount of kinetic energy and no potential energy. This maximum kinetic energy is equal in magnitude to the potential energy of the object due to its position just before it was released. This change of energy is an example of the **law of conservation of energy**. The law states that energy can never be created or destroyed; it is simply transferred from one form to the other.

Conservation Of Energy

Although energy is always conserved, no conversion of energy is ever 100 percent efficient. This means that useful energy output is always less than the input of energy. Some energy is wasted by such things as friction. For instance, not all the energy released from burning gasoline is converted into the motion of a car. In fact, a car engine is a very inefficient device because a large part of the energy transmitted through it is converted into heat. The energy from the heat and the motion together is equivalent to the energy from the gasoline. Therefore, energy is conserved, according to the law of conservation of energy. However, less energy is used to do work than is put into the system, and much useful energy is lost. The loss of useful energy is true for all energy transfers. The result is a decrease in the ability to do work.

Matter And Its Relation To Energy

Albert Einstein, one of the greatest physicists in history, discovered the relationship between energy and matter in his famous **theory of relativity**. According to Einstein, in any transformation, the total amount of matter and energy does not change. This law applies to everything we do from hammering a nail to launching a space shuttle.

The relationship between the quantity of matter and the amount of energy it stores is stated by Einstein's equation:

$$E = mc^2$$

E = the amount of energy m = the quantity of mass
c = the speed of light

This equation implies that energy cannot appear without a disappearance of mass. When gasoline is burned in a car engine, a chemical change causes one gram of gasoline to release 48,000 joules of energy. A **joule** is the metric system unit for energy. Almost all of this energy is released by the breaking of chemical bonds, but some matter is converted to energy as well. According to Einstein's equation, the amount of mass that disappears is 5.3×10^{-13} kilograms, or about half a billionth of a gram. If you could completely convert 1 ounce of matter into energy, you would obtain 700,000,000 kilowatt-hours of energy, enough to supply all the electric power New York City uses in a few weeks.

FURTHER EXPLORATION

Anything that can do work has energy. Electrical energy is one kind of energy. Nuclear energy is another kind. Gravitational potential energy is yet another kind. Work with a partner to name and research five different kinds of energy. Give examples of each kind of energy you research.

FUNTASTIC FACTS

■ The United States, the largest energy consumer in the world, uses about eight times more energy than its neighbor, Canada.

■ The first electrical generating station was established in New York City in 1882 by Thomas Edison.

WORD MEANINGS

● **kinetic**
kinetikos (GREEK)
to move

● **potential**
potens (LATIN)
to be able

TEMPERATURE

Why are the numbers 32, 0, and 273 really the same? They each are the temperature of the freezing point of water. People use three different scales to measure temperature: the Celsius scale, the Fahrenheit scale, and the Kelvin scale.

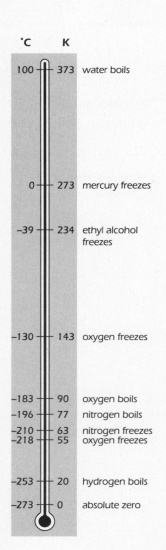

°C	K	
100	373	water boils
0	273	mercury freezes
−39	234	ethyl alcohol freezes
−130	143	oxygen freezes
−183	90	oxygen boils
−196	77	nitrogen boils
−210	63	nitrogen freezes
−218	55	oxygen freezes
−253	20	hydrogen boils
−273	0	absolute zero

Celsius And Fahrenheit

Most people measure temperature in degrees Celsius or Fahrenheit. **Celsius**, or **Centigrade**, is a metric system scale. Water freezes at 0°C and boils at 100°C. Celsius is used commonly in most countries except the United States. In the last few decades, however, Americans have begun to use Celsius more often.

Fahrenheit is part of the English system of measures and is most commonly used in the United States. Water freezes at 32°F and boils at 212°F.

The Kelvin Scale

Scientists and engineers frequently use the **Kelvin** temperature scale. Kelvin is also a metric system scale. A temperature in Kelvin is equal to the Celsius temperature plus 273.15. (The degree symbol is not used in Kelvin.) Kelvin is often more convenient than Celsius to use in scientific applications since many temperatures measured in the laboratory are very low and are written as negative numbers in Celsius. There are no negative temperature values on the Kelvin scale.

Absolute Zero

is the lowest temperature theoretically possible and is denoted as zero degrees Kelvin (0 K). Although this temperature has never been reached experimentally, scientists have determined this number by extrapolating a series of temperature readings. You might think that there is no energy at all at absolute zero - that the particles of matter stop moving completely. While the energy of particles may drop to a very low level, the particles of matter never actually stop moving.

$$K = °C + 273.15 \qquad °F = 9/5°C + 32$$
$$°C = K - 273.15 \qquad °C = 5/9°F - 32$$

SOUND

*I*f a tree falls in a forest and no one is there to hear it, has a sound been created? Some people might answer this question with a "no" because they believe sound is something that must be heard by a listener. Most scientists, however, would probably say, "yes".

What Is Sound?
Scientists might say that **sound** is a type of energy that is produced by vibrating matter. For us to hear a sound, it must be carried, or transmitted, through a medium. Solids, liquids, and gases are **mediums** that transmit sound. Solids are, in fact, the best transmitters of sound with the exception of carpets, drapes, and tiles. These solids absorb sound energy because they are made of materials that trap air.

Speed Of Sound
It takes time for sound to travel from place to place. As you move farther away from the source of the sound, this becomes apparent. For example, if you sit in a center field seat at a baseball game, the sound you hear after the batter hits the ball reaches your ears with a noticeable delay.

The speed of sound depends on three factors: **1.** the elasticity of gases, liquids, and solids, the three transmitters of sound. **2.** temperature of the air, and **3.** density. Sound travels faster in water than in ethyl alcohol, even though these two liquids may be at the same temperature. This is due to the fact that water has a greater density. Materials with greater density, higher temperatures, or greater elasticity increase the speed of sound.

The following chart will help you see how the speed of sound increases depending on the temperature of a gas (air), the elasticity of liquids and solids, and the density of liquids, solids, and gases.

FUNTASTIC FACTS

Sound cannot travel in a vacuum.

Sound travels about...
- 1100 feet per second in air.
- 4700 feet per second in water.
- 11,900 feet per second in brick.
- 16,400 feet per second in steel.

Material	Speed Of Sound (in miles per second)
Air at 0°C	331
Air at 20°C	344
Air at 100°C	390
Helium at 0°C	965
Ethyl alcohol at 25°C	1207
Water at 25°C	1498
Copper	3800
Wood (oak)	3850
Glass	5170
Steel	5200

Frequency

Sound frequency refers to the number of sound waves or compressions passing any given point per second. As the source vibrates faster, the frequency of the sound wave increases. The unit for frequency is the **hertz**, (named after Heinrich Hertz) and is also referred to as cycles per second, waves per second, or vibrations per second. The symbol for the hertz is Hz.

A frequency of one hertz means that one compression or wave vibrates per second; 100 Hz means that 100 compressions or waves vibrate per second.

Place a ruler on the end of a table and hold it down firmly. Hit the end of the ruler and take note of the sound produced as the ruler vibrates. Then shorten the length of the ruler by half and hit the ruler. Note that the sound is higher and that the ruler vibrates much faster. By shortening the length of the ruler and making it vibrate, you have increased the sound frequency. The higher the frequency, the higher the pitch; the lower the frequency, the lower the pitch. **Pitch** refers to the highness or lowness of a sound. Pitch should not be confused with the loudness or softness of a sound.

Amplitude

is the loudness or softness of a sound. A sound can be made louder or softer by changing the amount of energy applied to it. If you strike a tuning fork with a small amount of energy, it will produce a soft sound, or low amplitude. If you strike a tuning fork with more energy, you will produce a louder sound, or a sound with higher amplitude.

Regardless of the amount of energy used to produce a sound, or the amplitude of that sound, the nature of the sound itself remains the same. For example, gently pluck an E string on a guitar, and you will hear a soft E sound. If you pluck the string more vigorously, you will still hear the sound of E, but the amplitude will have increased.

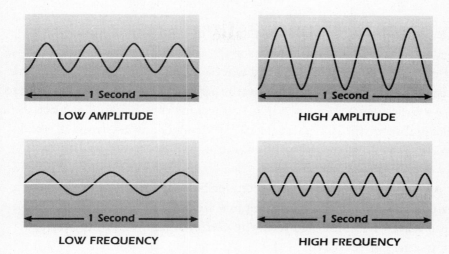

LOW AMPLITUDE

HIGH AMPLITUDE

LOW FREQUENCY

HIGH FREQUENCY

The more energy the sound wave has, the greater the amplitude, or the louder the sound. Electric amplifiers, used by many modern musicians, make sound louder by adding energy to the sound waves.

Sound Intensity
The amount of sound energy that reaches a given place or point per second is the **intensity** of the sound wave. Sound waves move out from the source in all directions. The farther you are from the source, the less intense the sound. The amplitude also decreases as sound waves move outward because some energy is continuously lost as it is transferred from particle to particle. One sound may seem louder than another of the same intensity if these sounds differ in pitch.

Sound levels are measured in units called **decibels**. A sound of 0 decibels is the quietest a human ear can detect. A decibel level of 140 is the pain threshold of the human ear. The following chart gives some decibel levels for comparison:

Sound	Sound Levels In Decibels (approximate)
Jet (30 meters away)	140
Music at a rock concert	120
Busy traffic	70
Conversation	65
Quiet car	50
Quiet radio	40
Whisper	20
Rustle of leaves	10

AM & FM Broadcasting

Radio waves travel at the speed of light (186,000 miles per second). Some waves travel along the surface of Earth and are called **ground waves**. Some waves travel up towards space and are reflected by a layer of the atmosphere called the **ionosphere**.

As radio waves move out in all directions they become weaker. A radio signal received 2 miles away is only one quarter as strong as a radio signal received 1 mile away. The same signal received 10 miles away is only one hundredth as strong.

A radio station is assigned a frequency called the **carrier frequency**, or **carrier signal**. Voice and music sounds are converted to electrical waves. These waves are combined with the carrier signal. They **modulate**, or change, the carrier wave.

Radio stations broadcast in either AM or FM mode. **AM** refers to amplitude modulation. In AM, the amplitude of the carrier wave is changed, not the frequency.

FM broadcasting refers to frequency modulation, or change in frequency. In FM, the sound frequency of voice and music is made to modulate the carrier wave.

AM and FM stations broadcast at very high carrier frequencies. For example, the unit used to describe AM broadcast signals is the kilohertz or KHz. One KHz is equal to 1000 Hz, or 1000 cycles per second. A radio station with a dial position of 880 broadcasts at a frequency of 880 KHz, or 880,000 Hz. This means that 880,000 waves arrive each second at the listener's radio. The unit used to describe FM broadcast signals is the megahertz or MHz. One MHz is equal to 1,000,000 hertz.

Harmonics

Harmonics are the components, or parts, of a musical tone. A **musical tone** is really a blend of several separate tones. This blend is referred to as a **composite tone**. A composite tone is a set of vibrations such as those made by a musical instrument or the human voice. For example, a guitar string vibrates over its entire length, but at the same time it also vibrates in sections called **partials**. Each partial is a fraction of the total length of the string. Each vibrating fraction, or partial, produces a separate tone, called an **overtone**.

The lowest tone in a composite sound is called the **fundamental**, or the first harmonic. This is produced by the vibration of the whole guitar. Other harmonics, or overtones, come from the vibrating partials. The second harmonic, called the **first overtone**, vibrates in two sections, so its frequencies are twice that of the fundamental, and therefore, an octave, or eight notes, higher.

Other overtones have frequencies three or more times that of the fundamental, depending on the fraction of the partial. For the most part, the higher overtones are made with less intensity (loudness) than lower ones. The fundamental plus the overtones form one complete tone that has the same pitch as the fundamental.

The number and intensity of a tone's harmonics determine the quality or **timbre** of the tone. When various musical instruments play the same note, it is the timbre that enables listeners to identify each instrument. In humans, harmonics create vowel sounds and the various tonal qualities that separate the sound of one voice from another.

Tuning A Guitar
One of the most accurate ways of tuning a guitar is by comparing a harmonic on one string to a harmonic on an adjacent string. By listening to and comparing harmonics, the listener can discern differences in pitch better than by comparing composite tones.

To play a composite tone on a guitar, you just pluck a string. To play a harmonic, you must use a special technique of gently touching a string with one hand while plucking it with the other.
A harmonic played on the guitar sounds somewhat like a bell.

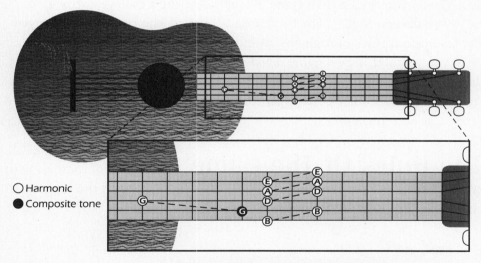

○ Harmonic
● Composite tone

TUNING A GUITAR BY COMPARISON OF HARMONICS
(Since there is no harmonic of G on the second string, a composite tone is used.)

FURTHER EXPLORATION
Discuss why you do not hear the sound of thunder until several seconds after seeing a bolt of lightning.

FOOTBALL

*I*n Europe, football is an obsession, a rough and tumble game watched by thousands of screaming fans, and televised to millions of contented European couch potatoes. On the other hand, football in the United States is an obsession, a rough and tumble game watched by thousands of screaming fans, and televised to millions of contented American couch potatoes. Of course, these two games are as different as day and night.

History Of The Game
Europeans refer to soccer as *football*. Americans play a completely different game. American football developed from the English game of rugby. During a college rugby game in 1874 between McGill University and Harvard, the two teams changed the traditional rules of rugby by running with the ball and throwing it. Over the years, this new game developed to become what football is to-day. Now it seems that almost everyone either plays football or watches it on TV. Usually, only college games and professional league games are broadcast on television.

The Rules Of The Game
The object of football is to cross the opponent's goal line with the ball. The team with the highest score wins the game.

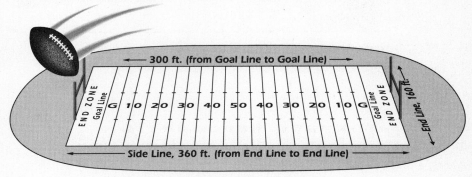

THE FOOTBALL PLAYING FIELD

Two opposing teams, each consisting of 11 players, may either run with the ball or pass it in a series of **plays**, or specific game strategies.

The game begins with one team **kicking off** to another. The ball is held on a device called a **tee**, and placed on the kicking team's 35-yard line. One or two **receivers** (players who try to catch the ball), from the other team are stationed on their own 5-yard line. Players attempt to tackle the opposition's receiver, (an **offensive** play), or protect their own receiver from the other team (a **defensive** play) so that the receiver can gain forward yardage.

If a ball that is kicked off travels more than ten yards, it becomes a **free ball**, and either team can recover it. The offense is allowed 4 plays, or **downs**, to try to advance the ball 10 yards by either running with the ball or passing it to another player. If the ball is moved forward 10 yards, the offense gets another first down. If the team fails to make a first down (10 yards in 4 plays), then the team must give up possession of the ball. The defense tries to prevent the offense from moving forward.

Time Of Play
Actual playing time is one hour. The game is divided into **halves** (30-minute periods), and each half is broken into **quarters** (15-minute periods). The second half begins with a kick-off; the second and fourth quarters begin play where it left off at the end of the previous quarter. An intermission called **half time** lasts about 20 minutes between halves.

The entire football game, in fact, usually lasts for several hours because the clock is stopped for a variety of time outs: time out for player injury, time out after a team scores, time out when the ball goes out of bounds, time out when the receiver fails to catch the ball, and so on.

FUNTASTIC FACTS

▪ **In the early days of football, players did not wear helmets or shoulder pads. Players carrying the ball wore straps so their teammates could pull them toward the goal line.**

▪ **The first Bowl game, the Rose Bowl, was played in 1902.**

━━ HOW TO SCORE IN FOOTBALL ━━

Points can be scored in a football game in the following ways:

❶ **touchdown**: A player crosses the goal line by running with or catching the football. (6 points)

❷ **conversion**: After a touchdown, the scoring team places the ball on the 20-yard line. The ball is then kicked through the goal post. (1 extra point)

❸ **field goal**: The ball is kicked through the goal post from anywhere on the field. (3 points)

❹ **safety**: This can be scored only by the defense and occurs when the ball carrier is tackled in his own endzone. (2 points)

FURTHER EXPLORATION

Time a "one hour" televised football game. Determine what percentage of broadcast time is used for commercials.